AF566911

Molecular Genetic Modification of Eucaryotes

Academic Press Rapid Manuscript Reproduction

Molecular Genetic Modification of Eucaryotes

edited by

Irwin Rubenstein
Department of Genetics and Cell Biology

Ronald L. Phillips
Charles E. Green
Department of Agronomy and Plant Genetics

Robert J. Desnick
Departments of Pediatrics and Genetics and Cell Biology

University of Minnesota
St. Paul, Minnesota

ACADEMIC PRESS New York San Francisco London 1977
A Subsidiary of Harcourt Brace Jovanovich, Publishers

ACADEMIC PRESS, INC.
111 Fifth Avenue, New York, New York 10003

United Kingdom Edition published by
ACADEMIC PRESS, INC. (LONDON) LTD.
24/28 Oval Road, London NW1

Library of Congress Cataloging in Publication Data

Molecular genetic modification of eucaryotes.

Proceedings of two workshops held at the University of Minnesota in 1974-75.
1. Genetic engineering–Congresses.
2. Molecular genetics–Congresses. I. Rubenstein, Irwin. II. Minnesota. University.

QH442.M65 575.12 77-13669
ISBN 0-12-601150-8

PRINTED IN THE UNITED STATES OF AMERICA

To
ALLEN S. FOX
1921–1977

Contents

Section II. GENETIC ASPECTS

Contributors

Numbers in parentheses indicate the pages on which the authors' contributions begin.

W. Adams, Jr. (13), *Department of Biology, Yale University, New Haven, Connecticut 06520*

Konrat Beyreuther (96), *Institüt für Genetics der Universität Koln, Germany*

F. Brenneman (13), *Department of Biology, Yale University, New Haven, Connecticut 06520*

Peter S. Carlson (43), *Department of Crop and Soil Sciences, Michigan State University, East Lansing, Michigan 48823*

F. Constabel (1), *National Research Council of Canada, Prairie Regional Laboratory, Saskatoon, Saskatchewan S7N 0W9, Canada*

P. R. Day (159), *Genetics Department, The Connecticut Agricultural Experiment Station, P.O. Box 1106, New Haven, Connecticut 06504*

Colin H. Doy (133), *Genetics Department, Research School of Biological Sciences, Australian National University, Canberra, Australia*

L. C. Fowke (1), *National Research Council of Canada, Prairie Regional Laboratory, Saskatoon, Saskatchewan S7N 0W9, Canada*

Allen S. Fox* (101), *Laboratory of Genetics, University of Wisconsin, Madison, Wisconsin 53706*

Y. Fuchs (13), *Division of Fruit and Vegetable Storage, The Volcani Center, P.O. Box 6, Bet Dagan, Israel*

A. W. Galston (13), *Department of Biology, Yale University, New Haven, Connecticut 06520*

* Deceased.

O. L. Gamborg (1), *National Research Council of Canada, Prairie Regional Laboratory, Saskatoon, Saskatchewan S7N 0W9, Canada*

Wolfgang Gerok (96), *Medizianische Universitäts Klinik, Freiburg, Germany*

Max E. Gottesman (77), *Section on Biochemical Genetics, Laboratory of Molecular Biology, National Cancer Institute, National Institutes of Health, Bethesda, Maryland 20014*

F. B. Holl (149), *National Research Council of Canada, Prairie Regional Laboratory, Saskatoon, Saskatchewan S7N 0W9, Canada*

Jurgen Horst (96), *Medizianische Universitäts Klinik, Freiburg, Germany*

K. N. Kao (1), *National Research Council of Canada, Prairie Regional Laboratory, Saskatoon, Saskatchewan S7N 0W9, Canada*

K. K. Kartha (1), *National Research Council of Canada, Prairie Regional Laboratory, Saskatoon, Saskatchewan S7N 0W9, Canada*

A. Kleinhofs (137), *Program in Genetics and Department of Agronomy and Soils, Washington State University, Pullman, Washington 99163*

Fridrich Kluge (96), *Medizianische Universitäts Klinik, Freiburg, Germany*

Carl R. Merril (83), *Laboratory of General and Comparative Biochemistry, National Institutes of Mental Health, Bethesda, Maryland 20014*

Oliver E. Nelson, Jr. (67), *Department of Genetics, University of Wisconsin, Madison, Wisconsin 53706*

K. Ohyama (1), *National Research Council of Canada, Prairie Regional Laboratory, Saskatoon, Saskatchewan S7N 0W9, Canada*

L. Pelcher (1), *National Research Council of Canada, Prairie Regional Laboratory, Saskatoon, Saskatchewan S7N 0W9, Canada*

M. Rancillac (13), *Institut National de la Recherche Agronomique, Station de Physiologie et Biochimie Végétales, "La Grande Ferrade," Centre de Recherches de Bordeaux, 33 Pont-de-la-Mage, Bordeaux, France*

R. K. Reid (13), *Department of Biology, Yale University, New Haven, Connecticut 06520*

Barry G. Rolfe (98), *Genetics Department, Research School of Biological Sciences, Australian National University, Canberra, Australia*

R. K. Sawhney (13), *Department of Biology, Yale University, New Haven, Connecticut 06520*

Kazunori Shimada (77), *Section on Biochemical Genetics, Laboratory of Molecular Biology, National Cancer Institute, National Institutes of Health, Bethesda, Maryland 20014*

B. Staskawicz (13), *School of Forestry and Environmental Studies, Yale University, New Haven, Connecticut 06520*

Robert A. Weisberg (77), *Laboratory of Molecular Genetics, National Institute of Child Health and Human Development, National Institutes of Health, Bethesda, Maryland 20014*

Jack M. Widholm (57), *Department of Agronomy, University of Illinois, Urbana, Illinois 61801*

Preface

Significant advances in molecular biology have been made since Watson and Crick described DNA and suggested its mode of replication. Certainly, molecular genetics has been one of the most exciting areas of science of this century. The worldwide food situation and the demands being placed on agriculture for increased production of high quality food have led to an intense interest in applying the scientific advances in molecular genetics, developed mostly with microbial systems, to plant and animal improvement for the betterment of mankind.

A transition appears to be occurring in the biological and agricultural sciences. In the biological sciences there is a shift from the investigation of model systems, principally procaryotic, to the study of applied systems such as economic crops. The intent is to apply the basic information generated over the past few decades to these higher plants. In the agricultural sciences, methods are being explored to reduce the complex economic crop plant to the cellular level. The intent is to manipulate and select for favorable phenotypes at the cellular level and then to regenerate complete plants that could be used in plant breeding programs. These transitions appear to be effecting a bridge that could result in a better utilization of the world's genetic resources. Researchers in the animal sciences might want to view these new approaches with plants as pilot systems that may provide future applications for the genetic engineering of animals. Indeed, the flow of information between plant and animal research is and will continue to enhance knowledge in the field.

With the foregoing ideas in mind, an interdisciplinary group of University of Minnesota faculty and graduate students commenced regular meetings in 1973. Members from eight departments in three colleges (Agriculture, Biological Sciences, and Health Sciences) participated in discussions on the molecular genetic modification of eucaryotes. Annual workshops on the topic have ensued, and this book represents the updated proceedings of the first two held in 1974 and 1975. The basic goals of the workshops were to review the current state of knowledge and techniques potentially useful for the molecular genetic modification of eucaryotes and to consider such questions as the following: What existing molecular biological techniques might be useful in carrying out genetic modification of eucaryotes? What techniques need to be developed? In what ways might molecular biological techniques be applied to plant improvement? Can techniques developed for plant improvement be applied to animal breeding and the treatment of human disease? What fundamental questions of cell biology and genetics need to be answered to facilitate the application of these techniques? How can these techniques be used to answer fundamental questions of cell biology and genetics? What are the main obstacles in the culture of plant cells and tissues to the successful application of molecular biological techniques? What are the advantages and disadvantages of the various plant materials presently available?

The editors hope that this book will serve to collect in one volume some independent judgments on the potential of molecular genetic modification, assess some of the tools at hand for such research, expose the more obviously deficient areas of knowledge, and present additional data. The common thread of the workshops, and of this book, was the interfacing of molecular genetics, plant cell and tissue culture, and plant improvement. The workshops were not widely publicized beyond the University of Minnesota and were attended by approximately 150 persons each year. It is our hope that this book will make available to the larger scientific community the philosophy and information presented at the workshops. We anticipate that the readers of this volume may have as diverse interests as those faculty members involved in the University of Minnesota "Molecular Genetic Modification Discussion Group" who organized the workshops. Thus the book will be of interest to many geneticists, cell biologists, plant breeders, plant physiologists, plant pathologists, and biochemists. Throughout the book the usefulness of molecular genetic techniques is obvious in three aspects: as an alternative approach to currently available genetic methods, as a new approach to currently unfeasible ideas, and as an approach to the study of the basic biology of the eucaryotic genetic system.

Papers presented at the workshops are integrated into two sections reflecting aspects of cell biology and genetics. Specific topics in the cellular aspects section include cell and tissue culture, protoplasts, somatic cell fusion, cellular mutagenesis, regeneration of new plant types, and the applicability of these techniques to plant improvements. Specific topics in the genetic aspects section include viruses and viral integration; integration and expression of foreign genetic material in human cells, *Drosophila*, plant cells, and legumes; biophysical studies of DNA uptake by plants; and genetic engineering for plant protection against diseases. Four presentations are not included in this volume due to the prior commitment of the authors to publish their papers, or similar papers having been published elsewhere; these include "Differentiation and morphogenesis of cell cultures into plants" by T. Murashige, "Techniques for the insertion of eucaryotic DNA into bacterial and viral DNA's—biochemical tools for genetic modification" by R. W. Davis, "Isolation of ovalbumin synthesizing polysomes—methods for the isolation of a gene" by R. T. Schimke, and "Molecular biology of nitrogen fixation" by R. C. Valentine.

The editors gratefully acknowledge financial support from the University of Minnesota departments of Agronomy and Plant Genetics, Biochemistry, Botany, Genetics and Cell Biology, Horticultural Science and Landscape Architecture, Microbiology, and Pediatrics, and from the University of Minnesota Educational Development Fund.

Protoplasts and Somatic Cell Hybridization in Plants

O.L. Gamborg, F. Constabel, K.N. Kao, L.C. Fowke, K. Ohyama,
L. Pelcher and K.K. Kartha

Somatic cell hybridization in plants by fusion has been the object of intense investigation because of the extraordinary potential of such a process. The interest has been motivated by the need to expand genetic accessability of desirable characteristics of plants and thereby accelerate progress in crop improvement (Witwer, 1974; Gamborg *et al*, 1974).

Somatic cell genetics and hybridization in seed plants have developed relatively slowly. A serious obstacle to rapid advances has been the lack of materials and a system that would be amenable to cellular genetic analyses and rapid cloning of desirable phenotypes. The use of whole plants requiring months to complete a life cycle makes genetic analysis a slow and tedious process. Somatic cells cultured *in vitro* may alleviate this problem.

In the last few years, significant advances have been made in the technology of plant cell culture (Street, 1973). Large populations of cells can be cultured under controlled environment in simple nutrient media on agar or in liquid suspension. These cells can be subjected to mutagenic treatments and plated under conditions for selection and cloning of desired genotypes.

Somatic plant cells possess the capacity for totipotency. Regeneration of complete plants from somatic cells has been reported for many species (Murashige, 1974). Recently, this technology has been extended to include plant protoplasts which are formed by means of enzymatic removal of the cell walls. The availability of protoplasts and a variety of tissue culture procedures make it possible to investigate somatic hybridization and other forms of genetic manipulation in higher plant cells (Gamborg and Wetter, 1975).

PROTOPLAST ISOLATION AND CULTURE

Procedures for the isolation and growth of protoplasts from cultured cells, leaves, and other organs are developing rapidly (Table 1). Protoplasts have been isolated from plant tissues by enzymatic removal of the cell walls (Fig. 1). Tissues or cells have been incubated in a solution containing osmotic stablizers, calcium salts, phosphate and a mixture of commercial enzyme preparations

TABLE 1
Examples of protoplasts in which cell regeneration and division has been observed.

Systematic	*Common Name*	*Cell Origin*
Ammi visnaga		culture
Bromus inermis	Brome grass	culture
Cicer arietinum	Chick pea	leaf
Brassica napus	Rapeseed	culture, leaf
Daucus carota	Carrot	culture, leaf
Glycine max	Soybean	culture
Linum usitatissimum	Flax	leaf, hypocotyl
Medicago sativa	Alfalfa	leaf, culture
Melilotus alba	Sweet clover	leaf
Phaseolus vulgaris	Bean	leaf, culture
Pisum sativum	Pea	leaf
Pisum sativum	Pea	culture, shoot tip
Vicia hajastana	–	culture
Vigna sinensis	Cow pea	leaf

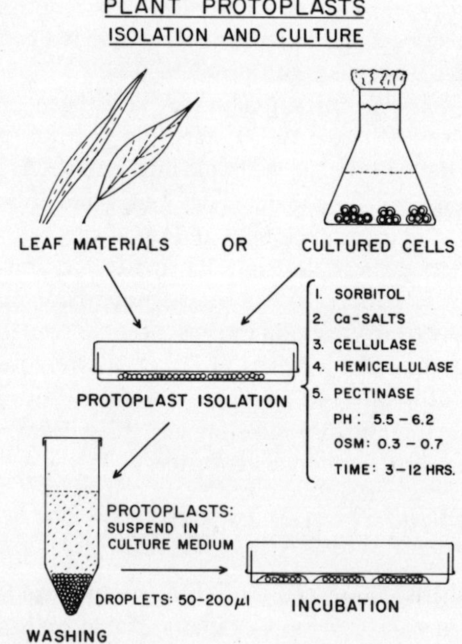

Fig. 1. Protoplast isolation and culture.

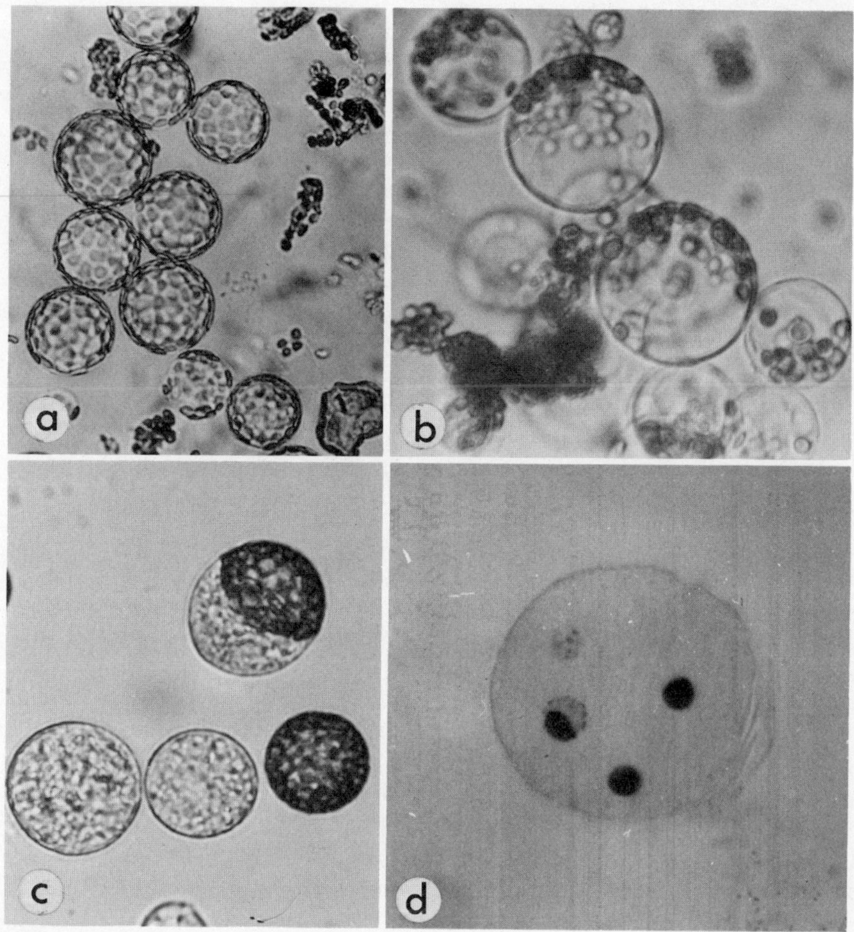

Fig. 2. a) Protoplasts from leaf tissue of rapeseed (Brassica napus). b) Protoplasts from corn leaf tissue. c) Protoplasts and a fusion product of carrot (light) and barley (dark) protoplasts. d) Heterokaryon showing nuclei of carrot (light) and barley (dark) and a nuclear fusion product (synkaryon).

which are predominantly cellulases, hemicellulases and pectinases (Constabel, 1975). After incubation for 3-12 hours, the debris was removed by filtration and the protoplasts were washed by centrifugation (Fig. 2, a, b).

Isolated protoplasts have been suspended in a nutrient medium and cultured as droplets in petri dishes (Kao *et al.*, 1971). The nutrient medium was similar to that used for plant tissue culture, but was modified to contain in addition sorbitol or similar osmotic stablizers. A suitable medium may consist of mineral salts, vitamins, glucose, ribose, casein hydrolyzate, 2, 4-dichlorophenoxyacetic acid, a cytokinin, and the osmotic stablizers. Other components, such as additional calcium salts, naphthaleneacetic acid and glutamine, may improve the survival rate and the possibility of cell wall regeneration and division.

A suitable pH may range from 5.5 to 6.2, and a suitable osmolality may vary from 0.3 to 0.7 depending upon the source of the protoplasts in culture. The isolated protoplasts were cultured as droplets with 10^4 -10^5 protoplasts per ml in petri dishes and incubated in humidified chambers at 25-28°C.

The cultured protoplasts have reformed the cell wall immediately after the enzymes were removed (Fowke *et al.*, 1974). The first division may occur within 1-3 days. After further incubation, cell clusters form which can be transferred to agar plates with similar nutritional and cultural conditions as used for droplets.

Uses of Plant Protoplasts

Plant protoplasts can be obtained in large quantities. They provide materials for new approaches to a variety of structural, physiological, biochemical and genetical problems. For example, the structure and biogenesis of membranes and cell walls might be studied more precisely and effectively with protoplasts than with other plant tissues. Cell wall removal has also permitted studies on membrane surface properties (Hartman *et al.*, 1973). The action of some plant pathogenic toxins has been elucidated through the use of protoplasts (Strobel, 1975). Protoplasts also have been employed successfully in the development of plant cell mutants and for uptake studies with viruses and DNA. Of all the attributes of protoplasts, perhaps the most remarkable has been their capacity for fusion.

Protoplast fusion

Spontaneous fusion between isolated protoplasts rarely occurs. Agents have been discovered, however, which promote fusion of protoplasts, apparently by increasing the area of membrane contact. The most successful techniques have employed alkaline-high calcium conditions (Keller and Melchers, 1973) or polyethylene glycol (PEG) (Kao *et al.*, 1974; Wallins *et al.*, 1974). The alkaline-high calcium method has been used in the fusion of tobacco protoplasts (Melchers and Labib, 1974), while PEG has proven effective in fusing protoplasts from a large number of plant taxa (Kao *et al.*, 1974).

Fusion has been initiated by adding a concentrated solution of PEG (M.W. 1540-6000) to protoplasts contained in a droplet placed in a petri dish. After 15-30 minutes, the PEG was gradually diluted out and replaced by culture medium. Fusion products were seen (Fig. 2, c). Unfused protoplasts and fusion products adhere to the surface of the petri dish. In most experiments, one of the protoplast species was obtained from leaf tissue and the other from cultured cells (Fig. 2, c).

Since the survival of unfused leaf protoplasts are low during culture, the heterokaryon fusion products could be recognized by the presence of green chloroplasts (Fig. 3, a). Up to 30% of the surviving protoplasts may be heterokaryons. The heterokaryons contained different proportions of nuclei from the two protoplast species (Constable *et al.*, 1975a). Large heterokaryons arising by multiple fusions and containing several nuclei often failed to survive. A proportion of the heterokaryons regenerate a cell wall and divide (Kao *et al.*, 1974; Constable *et al.*, 1975b; Kartha *et al.*, 1974b). The fusion of nuclei from the two parental species may occur during mitosis in the heterokaryon (Kao *et al.*, 1974). Recent results support this, indicating that interphase nuclei of two different species within the heterokaryon can fuse (Constable *et al.*, 1975a; Dudits *et al.*, 1976) (Fig. 2, d). The process may be initiated by the formation of nuclear membrane bridges (Fowke *et al.*, 1975). It has not been possible to establish that such hybrid nuclei enter mitosis.

Division has been observed in heterokaryons encompassing a wide range of plant genera and families (See Table 2) and there has been no indication of cellular incompatibility (Fig. 3, b). Fusion combinations such as pea x soybean and corn x soybean have yielded fusion products where both sets of parental chromosomes have been identified in metaphase plates (Kao *et al.*, 1974; Constable *et al.*, 1975a). In the systems where heterokaryons divided and formed cell clusters, the hybrids could be recognized visually. Cells with appropriate phenotypic markers for selection were required to isolate hybrid clones (Constable *et al.*, 1975b).

Uptake studies

Various reports describe the uptake by protoplasts of nuclei, chloroplasts, and DNA. Bonnet and Eriksson (1974) have investigated the uptake of algal chloroplasts by protoplasts from carrot cell cultures. They employed PEG to facilitate the process. Uptake consistantly occurred in the presence of 28% PEG (Modepeg, M.W. 1500). Up to 16% of the viable protoplasts contained one or more chloroplasts which entered the cytoplasm. The PEG activated the plasma membrane but did not appear to affect the chloroplast membranes (Bonnet and Eriksson, 1974). Although carrot protoplasts tolerated the PEG treatment, the fate of the algal chloroplasts was not ascertained.

TABLE 2.
*Examples of plant genera in which protoplast fusion and heterokaryocyte division has been observed.**

SOURCE OF PROTOPLASTS		
Leaf mesophyll		*Cell Culture*
Barley *(Hordeum vulgare)*	x	soybean *(Glycine max)*
Pea *(Pisum sativum)*	x	*Vicia hajastana*
Corn *(Zea mays)*	x	soybean *(Glycine max)*
Pea *(Pisum sativum)*	x	soybean *(Glycine max)*
Rapeseed *(Brassica napus)*	x	soybean *(Glycine max)*
Alfalfa *(Medicago sativa)*	x	soybean *(Glycine max)*
Sweet clover *(Melilotus alba)*	x	soybean *(Glycine max)*
Chick pea *(Cicer arietinum)*	x	soybean *(Glycine max)*
Angelica archangelica	x	carrot *(Daucus carota)*

*References
(Kao and Michayluk, 1974; Kao *et al.*, 1974; Kartha *et al.*, 1974b; Fowke *et al.*, 1975; Dudits *et al.*, 1976; Constable *et al.*, 1975a, 1975b).

Uptake of bacterial DNA has been reported in protoplasts of several plant species (Ohyama, *et al.*, 1972). This DNA appeared to be taken up in the double-stranded form. Uptake was enhanced by polycations. This bacterial DNA in protoplasts was subject to degradation, the extent of which varied between plant species. Using homologous DNA, Hoffman and Hess (1973) observed that DNA taken up by petunia protoplasts became associated with the nucleus.

The success of transformation in bacterial cells has been related to the absence of particular nucleases (Heyn *et al.*, 1974). Conceivably, plants may vary with respect to enzymes which recognize and degrade non-homologous DNA. The information on plant DNase is scant. A comprehensive study of these enzymes, and of DNA uptake and incorporation mechanisms would be extremely valuable (Holl, 1973).

The uptake of DNA and genetic transformation also has been investigated with seeds and other plant organs (Holl *et al.*, 1974). These systems have the advantage that complete plant formation, after DNA uptake, may be achieved.

The uptake of nuclei has been observed in petunia protoplasts (Potrykus and Hoffman, 1973). In the future organelle uptake may be extended to metaphase chromosomes, a method which has been employed successfully in mammalian systems (Willecke and Ruddle, 1975).

The choice of plant systems for genetic transformation studies by DNA or organelle uptake has been restricted. Phenotypic markers detectable at the cellular level which would permit effective selection and cloning are not yet available.

Selection systems

Somatic hybridization in mammalian cells has been greatly aided by drug-resistant mutants (Littlefield and Goldstein, 1970; Davidson, 1974). One cell line is resistant to 5-bromodeoxyuridine (BUdR) and its cells lack thymidine kinase (TK). Another mutant cell line is resistant to 8-azaguanine and it lacks hypoxanthine-guanine phosphoribosyl transferase. When cells of these two lines were fused, the hybrids were isolated in a selective medium containing hypoxanthine, aminopterin and thymidine. Neither of the parental lines survived since aminopterin blocks *de novo* synthesis of thymidine monophosphate and the exogenous compounds could not be utilized. Hybrids contained both enzymes and grew in the presence of aminopterin. Mutants of this type would be particularly suitable for plant protoplast fusion studies because a positive selection system for the hybrids could then be employed.

BUdR-resistant plant cell mutants have been reported (Chaleff and Carlson, 1974). A similar resistant line from soybean suspension cultures also has been isolated and studied (Ohyama, 1974). Cells of a more recent BUdR-resistant clone grow at a very high rate in the presence of 1 mg/ml BUdR. In contrast to the BUdR-resistant mammalian cells, the soybean mutant contained active thymidine kinase and incorporated the BUdR into DNA. The cells retained resistance during culture in the absence of BUdR. The metabolic basis for the resistance has not been entirely clarified.

A number of plant tissue cultures with resistance to natural amino acids and analogs of amino acids have been reported (Chaleff and Carlson, 1974; Green and Phillips, 1974; Widholm, 1974). Until the biochemical basis for the resistance has been fully established, it is not possible to predict the usefulness of these mutants in selection procedures.

Some toxins produced by plant pathogens exhibit a high degree of specificity. For example, the toxins from *Helminthosporium maydis* race T selectively destroyed protoplasts from susceptible but not resistant lines of maize (Pelcher *et al.*, 1975). In a fusion experiment, perhaps hybrid cells could be selected by using appropriate cell lines and a combination of toxins and antimetabolites. Such a system would have potential of a practical nature, assuming toxin-resistant lines could regenerate plants which retained resistance to the pathogen. Albino and light-sensitive mutants of species where plant regeneration can occur from protoplasts provide an effective hybrid selection system (Melchers and Labib, 1974).

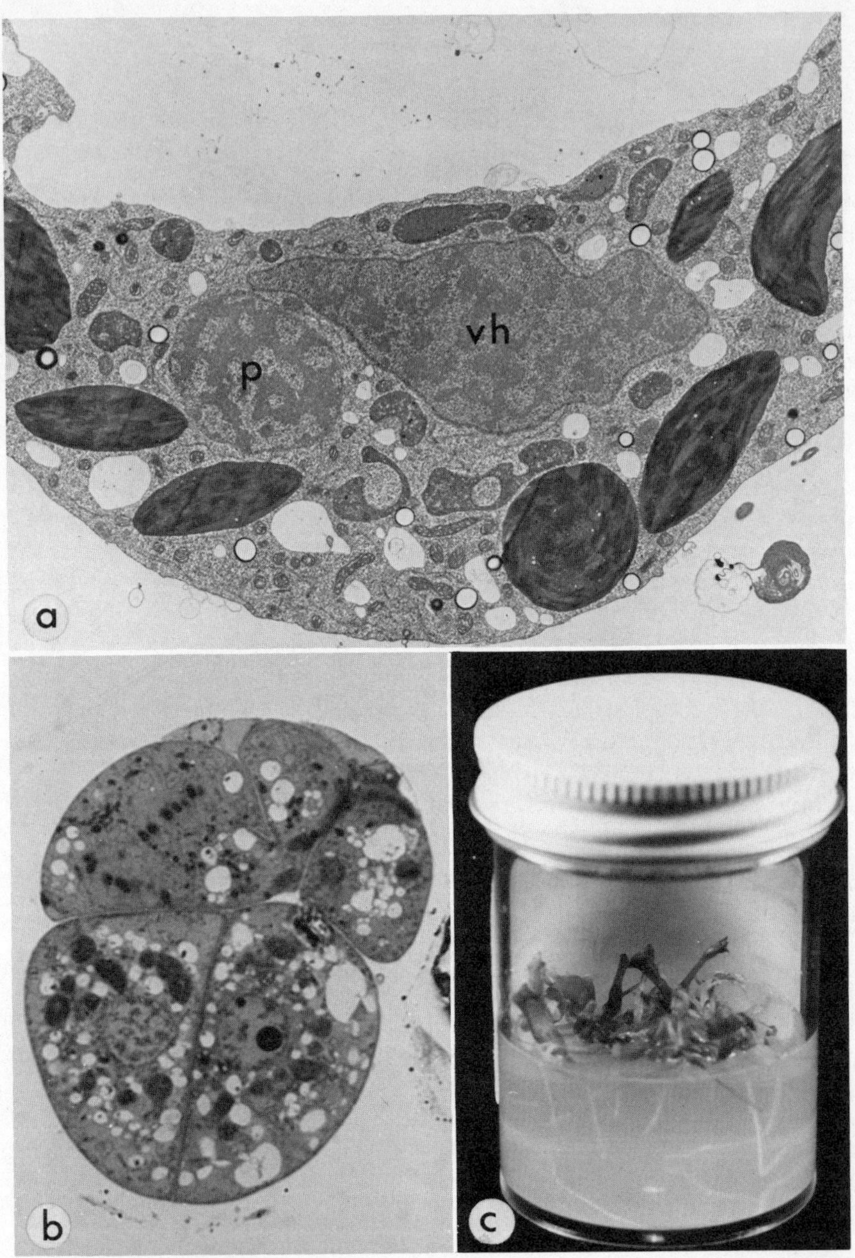

Fig. 3. a) Section of fusion product showing nucleus and chloroplasts from pea leaf (p) and nucleus from Vicia hajastana (vh). b) Cell hybrids of soybean and sweet clover. c) Plantlets obtained from carrot protoplasts by embryogenesis.

PLANT REGENERATION

Plant cell cultures derived from shoots, embryos and other organs have the potential to differentiate organized structures and complete plants. The actual capacity for expressing morphogenesis may vary greatly between plant families (Kartha *et al.*, 1974a; Gamborg *et al.*, 1975; Green and Phillips, 1975; Kartha *et al.*, 1976). In most species of agronomic importance, however, plant regeneration from cells grown in liquid shake cultures has not been achieved. Protoplasts also have the potential to differentiate and form complete plants (Grambow *et al.*, 1972; Kartha *et al.*, 1974). Plants have been obtained from cells of fused protoplasts of two tobacco species (Carlson *et al.*, 1972; Melchers and Labib, 1974).

In carrot and a few other species, cultured cells have developed into embryos, a process which has not required exogenous hormones. Embryogenesis also has occurred in protoplasts from carrot cell cultures (Fig. 3, c). For this reason, protoplasts from carrot and other species with embryogenically active cells may provide valuable materials for fusion and hybrid plant regeneration studies. (Dudits *et al.*, 1976). Other desirable protoplast sources may include tissues which have been induced to differentiate and regenerate plants. Conceivably, the protoplasts from such tissues may elicit the morphogenetic response after fusion. Morphogenesis in intergeneric somatic cell hybrids has not been reported, but as appropriate selection and culture methods become available, plant regeneration becomes increasingly feasible.

CONCLUSIONS

Significant progress has been made in the technology of tissue culture and somatic cell hybridization in higher plants. Protoplast isolation and successful culture has been realized in an increasing number of plant species. Fusion has been achieved routinely at high rates with protoplasts from different plant genera and families. Heterokaryons have divided and nuclear fusion has been demonstrated. The hybrid cells have been recognized by the presence of phenotypic markers such as chloroplasts. Their true hybrid nature has been demonstrated by nuclear staining and ultrastructural analysis. There has been no apparent evidence of incompatibility in the fusion products or the hybrid cells.

Exploratory research has been initiated on exchange of genetic information by chloroplast and nuclear transplantation into protoplasts. Plants have been regenerated from protoplasts of a limited number of species and from fusion products of tobacco. This technology will likely be extended and may include the regeneration of complete plants from intergeneric hybrid cells.

The tissue culture technology with somatic hybridization could provide methodology to supplement conventional plant breeding procedures. The ability to circumvent the sexual cycle might permit hybridization between normally incompatible plant genera. An extensive range of wide crosses can be envisaged. Moreover, the technology also would have a particular advantage for improvement of root and tuber crops which are propagated vegetatively. Some of these crops do not go through the sexual cycle. By employing somatic hybridization, desirable crosses may be achieved.

Existing tissue culture methods which employ mutagenesis and selection can be used to obtain genetic variability, but somatic hybridization techniques could expand the genetic base far beyond what is accessible to the plant breeder at present. There is some skepticism about the feasibility and practical use of the developing technology. The continued urgency, however, for new crops with superior productivity, growth efficiency, and product quality constitutes a vital and compelling reason for attempting to establish a practical technology for producing intergeneric somatic hybrid plants.

REFERENCES

Bonnett, H.T., and Eriksson, T. (1974). Transfer of algal chloroplasts into protoplasts of higher plants. *Planta 120,* 71-79.

Carlson, P.S., Smith, H.H., and Dearing, R.D. (1972). Parasexual interspecific plant hybridization. *Proc. Nat. Acad. Sci. (U.S.A.) 69,* 2292-2294.

Chaleff, R.S., and Carlson, P.S. (1974). Somatic cell genetics of higher plants. *Ann. Rev. Genetics. 8,* 267-278.

Constabel, F. (1975). Protoplast isolation and culture. *In Plant Tissue Culture Methods.* (O.L. Gamborg and L.R. Wetters, eds.), pp. 11-21. National Research Council of Canada, Saskatoon.

Constabel, F., Dudits, D., Gamborg, O.L., and Kao, K.N. (1975a). Nuclear fusion in intergeneric heterokaryons. *Can. J. Bot. 53,* 2092-2095.

Constabel, F., Kirkpatrick, J.W., Kao, K.N., and Kartha, K.K. (1975b). The effect of canavanine on the growth of cells from suspension cultures and on intergeneric heterokaryons of canavanine sensitive and tolerant plants. *Biochem. Physiol. Pflanzen. 168,* 319-325.

Davidson, R.L. (1974). Gene Expression in somatic cell hybrids. *Ann. Rev. Genetics. 8,* 195-218.

Dudits, D., Kao, K.N., Constabel, F., and Gamborg, O.L. (1976). Fusion of carrot and barley protoplasts and division of heterokaryocytes. *Exp. Cell Res. In press.*

Fowke, L. C., Bech-Hansen, C.W., Constabel, F., and Gamborg, O.L. (1974). A comparative study on the ultrastructure of cultured cells and protoplasts of soybean during cell division. *Protoplasma 81,* 189-203.

Fowke, L.C., Bech-Hansen, C.W., Gamborg, O.L., and Constabel, F. (1975). Electron microscopic observations of mitosis and cytokinesis in multinucleate protoplasts of soybean. *J. Cell Sci. 18,* 491-507.

Gamborg, O.L., Constabel, F., Fowke, L.C., Kao, K.N., Ohyama, K., Kartha, K.K., and Pelcher, L. (1974). Protoplast and cell culture methods in somatic hybridization in higher plants. *Can. J. Genet. Cytol. 16,* 737-750.

Gamborg, O.L., and Wetter, L.R. (1975). *Plant Tissue Culture Methods.* National Research Council of Canada, Saskatoon.

Gamborg, O.L., Shyluk, J., and Kartha, K.K., (1975). Factors affecting the isolation and callus formation in protoplasts from the shoot apices of *Pisum sativum* L. *Plant Science Letters 4,* 285-292.

Grambow, H.J., Kao, K.N., Miller, R.A., and Gamborg, O.L. (1972). Cell division and plant development from protoplasts of carrot cell suspension cultures. *Planta 103,* 348-355.

Green, C.E., and Phillips, R.L. (1974). Potential selection system for mutants with increased lysine, threonine and methionine in cereal crops. *Crop Sci. 14,* 827-830.

Green, C.E., and Phillips, R.L. (1975). Plant regeneration from tissue cultures of maize. *Crop Sci. 15,* 417-421.

Hartmann, J.X., Kao, K.N., Gamborg, O.L., and Miller, R.A. (1973). Immunological methods for agglutination of protoplasts from cell suspension cultures of different genera. *Planta 112,* 45-56.

Heyn, R.F., Rorsch, A., and Schilperoort, R.A. (1974). Prospects in genetic engineering of plants. *Quart. Rev. Biophys. 7,* 35-73.

Hoffman, F., and Hess, D. (1973). Uptake of redioactively labelled DNA into isolated protoplasts of *Petunia hybrida. A. Pflanzenphysiol. 69,* 81-83.

Holl, F.B. (1973). Cellular environment and the transfer of genetic information. *Colloq. Int. Cent. Nat. Rech. Sci. 212,* 509-516.

Holl, F.B., Gamborg, O.L., Ohyama, K., and Pelcher, L.E. (1974). Genetic transformation in plants. *In Tissue Culture and Plant Science 1974.* (H.E. Street, ed.), pp. 301-327. Academic Press, New York.

Kao, K.N., Gamborg, O.L., Miller, R.A., and Keller, W.A. (1971). Cell divisions in cells regenerated from protoplasts of soybean and *Haplopappus gracilis. Nature New Biology 232,* 124.

Kao, K.N., and Michayluk, M.R. (1974). A method for high frequency intergeneric fusion of plant protoplasts. *Planta 115,* 355-367.

Kao, K.N., Constabel, F., Michayluk, M.R., and Gamborg, O.L. (1974). Plant protoplast fusion and growth of intergeneric hybrid cells. *Planta 120,* 215-227.

Kartha, K.K., Michayluk, M.R., Kao, K.N., Gamborg, O.L., and Constabel, F. (1974a). Callus formation and plant regeneration from mesophyll protoplasts of rape plants *(Brassica napus* L. cv. Zephyr). *Plant Science Letters 3,* 265-271.

Kartha, K.K., Gamborg, O.L., Constabel, F., and Kao, K.N. (1974b). Fusion of rapeseed and soybean protoplasts and subsequent division of heterokaryocytes. *Can. J. Bot. 52,* 2435-2436.

Kartha, K.K., Gamborg, O.L., Shyluk, J.P., and Constabel, F. (1976). Morphogenetic investigations on *in vitro* leaf culture of tomato *(Lycopersicon esculentum* Mill. cv. Starfire) and high frequency plant regeneration. *Z. Pflanzenphysiol. 77,* 292-301.

Keller, W.A., and Melchers, G. (1973). The effect of high pH and calcium on tobacco leaf protoplast fusion. *Z. Naturforsch. C28,* 737-741.

Littlefield, J.W., and Goldstein, S. (1970). Some aspect of somatic cell hybridization. *In Vitro 6,* 21-31.

Melchers, G., and Labib, G. (1974). Somatic hybridization of plants by fusion of protoplasts. I. Selection of light resistant hybrids of "haploid" light sensitive varieties of tobacco. *Molec. Gen. Genet. 135,* 277-294.

Murashige, T. (1974). Plant propagation through tissue cultures. *Ann. Rev. Plant Physiol. 25,* 135-166.

Ohyama, K. (1974). Properties of 5-bromodeoxyuridine-resistant lines of higher plant cells in liquid culture. *Exp. Cell. Res. 89,* 31-38.

Ohyama, K., Gamborg, O.L., and Miller, R.A. (1972). Uptake of exogenous DNA by plant protoplasts. *Can. J. Bot. 50,* 2077-2080.

Pelcher, L.E., Kao, K.N., Gamborg, O.L., Yoder, O.C., and Gracen, V.E. (1975). Effects of *Helminthosporium maydis* race T toxin on protoplasts of resistant and susceptible corn *(Zea mays* L.). *Can. J. Bot. 53,* 427-431.

Potrykus, I., and Hoffman, F. (1973). Transplantation of nuclei into protoplasts of higher plants. *Z. Pflanzenphysiol. 69,* 287-289.

Street, H.E. (1973). *Plant Tissue and Cell Culture* Botanical Monographs Vol. II. Blackwell Scientific Publications, London.

Strobel, G.A. (1975). A mechanism of disease resistance in plants. *Sci. Amer. 232(l),* 80-89.

Wallins, A., Glimelius, K.G., and Eriksson, T. (1974). The induction of aggregation and fusion of *Daucus carota* protoplasts by polyethylene glycol. *Z. Pflanzenphysiol. 74,* 64-80.

Widholm, J.M. (1974). Selection and characteristics of biochemical mutants of cultured plant cells. *In Tissue Culture and Plant Science 1974.* H.E. Street (ed.), pp. 287-300. Academic Press, New York.

Willecke, K., and Ruddle, F.H. (1975). Transfer of the human gene for hypoxanthine-guanine phosporibosyl-transferase via isolated human metaphase chromosomes into mouse L. Cells. *Proc. Nat. Acad. Sci. (U.S.A.). 72,* 1792-1796.

Witwer, S.H. (1974). Maximum production capacity of food crops. *BioSci. 24,* 216-224.

Opportunities and Obstacles in the Culture of Cereal Protoplasts and Calluses.

A. W. Galston, W. Adams, Jr., F. Brenneman, Y. Fuchs, M. Rancillac, R. K. Reid, R. K. Sawhney and B. Staskawicz

Several developments of the last fifteen years have led plant physiologists to the belief that they may be able to improve important crop plants significantly through the use of techniques based on isolated protoplasts. Among the new developments have been successful preparation of masses of plant protoplasts by enzymatic treatment of tissues (Cocking, 1960), the demonstration that individual tobacco protoplasts (Takebe *et al.*, 1971) or their somatically fused hybrid products (Carlson *et al.*, 1972; Melchers and Labib, 1974) can regenerate entire plants, and reports of successful transduction (Doy *et al.*, 1973; Johnson *et al.*, 1973) and transformation (Hess, 1973; Ledoux *et al.*, 1974) of plant cells. So far, protoplast technology has not yielded any agronomically important results, at least partly because the important agricultural plants, like the cereals and legumes, do not readily progress from isolated protoplast to callus or from undifferentiated callus to plants.

In the hope that protoplast cultivation and the techniques of molecular biology might be adapted for the production of new types of cereals and legumes, we have undertaken a systematic study of procedures for the isolation of protoplasts, regeneration of cell walls, induction of subsequent cell division and the possible regeneration of whole plants. With oats, described in this report, only limited and occasional cell division by the cells regenerated from protoplasts have so far been induced. We have therefore also turned to studies of the growth of oat callus in an attempt to induce regeneration of whole plants.

In connection with biochemical evaluation of isolated protoplasts, we have studied their ability to incorporate ^{3}H-labelled leucine, uridine and thymidine into protein, RNA and DNA, respectively. We have also investigated the effects of hormones, inhibitors, and the phytotoxin victorin on macromolecular synthesis in both resistant and susceptible oat varieties. Finally, as a prelude to attempted transformation of resistant to sensitive varieties (or vice versa), we have developed an efficient procedure for the extraction and purification of native DNA from oat leaves.

PREVIOUS WORK ON CULTURE OF MONOCOTYLEDONOUS CALLUSES, CELLS, AND PROTOPLASTS

Successful culture of monocotyledonous plant tissues was achieved much later than the first success with dicotyledonous plants, but now includes both cereals and non-cereals. In the latter group are included asparagus (Loo, 1946; Galston, 1948; Wilmar and Hellendoorn, 1968; Jullien, 1973a, b,), orchids (Churchill *et al.*, 1973), lily (Sheridan, 1968), gladiolus (Ziv *et al.*, 1969; Simonsen and Hildebrandt, 1971); onion (Fridborg, 1971), garlic (Havranek and Novak, 1972), coconut (Blake *et al.*, 1974), sugarcane (Nickell and Maretzki, 1969, 1970) and brome grass (Gamborg *et al.*, 1970).

Cereal tissue cultures include maize endosperm (Straus, 1954, 1960), rye endosperm (Norstog, 1956), rye embryo callus (Carew and Schwarting, 1958), and corn callus from stems and roots (Mascarenhas *et al.*, 1965; Hendre *et al.*, 1972; Sheridan, 1972). Since 1967, rice has been successfully cultured, both on undefined and chemically defined media (Yatazawa *et al.*, 1967; Yamada *et al.*, 1967; Tamura, 1968; Kawata and Ishihara, 1968), and highly productive suspension cultures have been achieved (Ohira *et al.*, 1973; Lieb *et al.*, 1973). Wheat callus has been successfully cultured (Trione, 1968; Gamborg and Eveleigh, 1968; Warick and Fuchs, 1968; Shimada *et al.*, 1969) and its nitrogen requirements studied (Gamborg, 1970; Bayley *et al.*, 1972). Other successes include sorghum (Masteller and Holden, 1970; Hendre *et al.*, 1972; Schenk and Hildebrandt, 1972), barley callus (Gamborg and Eveleigh, 1968; Schenk and Hildebrandt, 1972) and oat callus (Carter *et al.*, 1967).

Success in organ differentiation has been achieved in asparagus (Galston, 1948; Wilmar and Hellendoorn, 1968; Aynsley, 1974), lily (Sheridan, 1968; Hussey, 1974), garlic (Fridborg, 1971; Havranek and Novak, 1973; Fridborg and Eriksson, 1974), gladiolus (Ziv *et al.*, 1970; Simonsen and Hildebrandt, 1971), onion (Mackenzie *et al.*, 1974), sugarcane (Heinz and Mee, 1969; Barba and Nickell, 1969) and brome grass (Gamborg *et al.*, 1970). Less dramatic results have been obtained with cereals, although root and shoot formation have been noted with callus cultures of rice (Tamura, 1968; Kawata and Ishihara, 1968; Nishi *et al.*, 1968) and sorghum (Masteller and Holden, 1970) transfered from a medium containing, to one lacking 2, 4-dichlorophenoxyacetic acid (2, 4-D). This also seems to work in barley callus (Cheng and Smith, 1974), oat callus (Carter *et al.*, 1967; Cummings *et al.*,, 1976), wheat callus (Trione *et al.*, 1968; Shimada *et al.*, 1969; Prokhorov *et al.*, 1974), and corn callus (Green and Phillips, 1975).

Anther culture has been used to produce haploid calluses in a few cereals. The concentrations of indole-3 acetic acid (IAA), kinetin and 2, 4-D were critical in rice (Niizeki and Oono, 1968; Iyer and Raina, 1972) and barley (Bouharmont *et al.*, 1974). Wheat anther callus gave rise to plantlets when 2, 4-D was excluded from the medium and IAA or 2-napthalene acetic acid (NAA) and kinetin were included (Picard and Buyser, 1973; Ouyang *et al.*, 1973; Wang *et al.*, 1973). Root differentiation from callus derived from corn anther culture was reported by Murakami and co-workers (1973).

Although protoplasts have been made to grow into entire plants in about half a dozen dicotyledonous plants, success with monocotyledons has been limited to two species, brome grass (Kao *et al.*, 1973) and asparagus (Bui-Dang-Ha and Mackenzie, 1973). In addition, cell division and callus formation without plant regeneration has been accomplished with sugarcane protoplasts (Maretzki and Nickell, 1973).

Cereal protoplasts were studied starting in 1965, when Ruesink and Thimann examined the behavior of membranes of oat coleoptile protoplasts subjected to enzymes, detergents, and IAA (Ruesink and Thimann, 1965; Ruesink, 1971). Later, Taiz and Jones (1971) prepared barley aleurone protoplasts by enzymatic techniques, Wakasa (1973) reported the preparation of protoplasts from a number of cereals and Evans and co-workers (1972) reported the preparation of wheat, rye, oat, and barley protoplasts from mature leaves and their culture on a modified White's medium. Wheat and rye protoplasts readily formed cell walls; oat and barley were less successful. Recently, Vasil and Vasil (1974) prepared corn root and leaf protoplasts, and cultured them on media conducive to cell wall synthesis. The protoplasts from roots did not divide, a few from leaves divided once.

The possibilities for the combination of genetic material from otherwise incompatible plants has prompted an investigation into somatic fusion of protoplasts. Power and co-workers (1970) and Cocking (1973) reported fusion of oats, corn, wheat and barley protoplasts in the presence of $NaNO_3$. Giles (1974) observed corn protoplast fusion, and Fodil and co-workers (1971) reported fusion of oat protoplasts. The fascination of a possible nitrogen-fixing cereal plant has initiated investigations into the fusion of cereal protoplasts with those of a legume. Successful fusion of soybean and wheat protoplasts has been reported (Gamborg *et al.*, 1972), and the nuclei of the resulting heterocaryon divide synchronously. Recently soybean and barley protoplasts were fused in the presence of polyethylene glycol and high concentrations of calcium. The fusion product divided and formed callus (Constabel and Kao, 1974).

Clearly, the possibilities for plant improvement by way of protoplast manipulation rest on the development of reproducible methods for regeneration of entire plants. Such methods are especially unsatisfactory in the cereals, the object of these investigations.

CULTURE OF OAT PROTOPLASTS

Preparation of protoplasts

Seedlings of *Avena sativa* L. varieties Garry and Victory were grown from seed in vermiculite and subirrigated twice a day with a commercial salt solution ("Hyponex", 7-6-19 analysis, 1.2 g/l). Growth proceeded in environmental chambers at 24° C with 16 hours of light per day (fluorescent cool white VHO bulbs supplemented with 40 W incandescent bulbs, 9:1 energy ratio, providing approximately 12,000 lux).

The seedlings were grown for 6 to 7 days until they were about 10 cm high. The leaves were cut 7 cm from the tip and, while still turgid, were sterilized for 60 seconds in 70% ethanol, washed twice with sterile water and blotted. From this point, all procedures were aseptic. The lower epidermis of the leaves was peeled off, and the tips and damaged material removed. The leaves were placed peeled side down on 3 ml of a cellulase solution in 30 ml beakers. Where Onozuka SS cellulase (All JapanBiochemicals) was used, the leaves were incubated for 18 hours at 24° C in the dark on a 2% solution of the enzyme in B5 "A" salts (Grambow *et al.*, 1972). When cellulysin (Cal. Biochem) was used, incubation was for 2 hours at 24° C at a concentration of 1%. The leaf "skeletons" were agitated gently to remove any remaining protoplasts and were then discarded. The suspensions of protoplasts contained some chloroplasts, but no other visible debris. The yield was of the order of 9×10^6 protoplasts per gram wet leaf tissue, as determined in a hemacytometer.

The suspensions were centrifuged at 15 g for 20 minutes to pellet the protoplasts. The supernatant solutions were removed and the pellets were washed twice with 5 ml of B5 "A" salts (Grambow *et al.*, 1972). Viability of the protoplasts was determined by their uptake of neutral red (Stadelmann and Kinzel, 1972), or rejection of erythrosin B (Merchant *et al.*, 1964).

Plating of Protoplasts

After the final centrifugation, the pellets were suspended in 2 ml of B5 "B" medium (Grambow *et al.*, 1972) and mixed with equal volumes of "B" medium containing 1.2% Ionagar (Colab Laboratories, Inc., Illinois) at a temperature of 45° C. The suspensions were poured into glass petri dishes sealed with Parafilm and incubated in environmental growth chambers at 22° C in full light (18 hours/day), in diminished intensity provided by several layers of cheesecloth, or

Fig. 1. Development of oat leaf protoplasts cultured in agar medium.
- *a. Protoplasts at onset of incubation.*
- *b. Protoplasts enlarging.*
- *c-d. Enlarged protoplasts with thinning of chloroplasts.*
- *e-f. Enlarged cells with walls.*
- *g-j. Doubles, triples, and clusters of cells.*

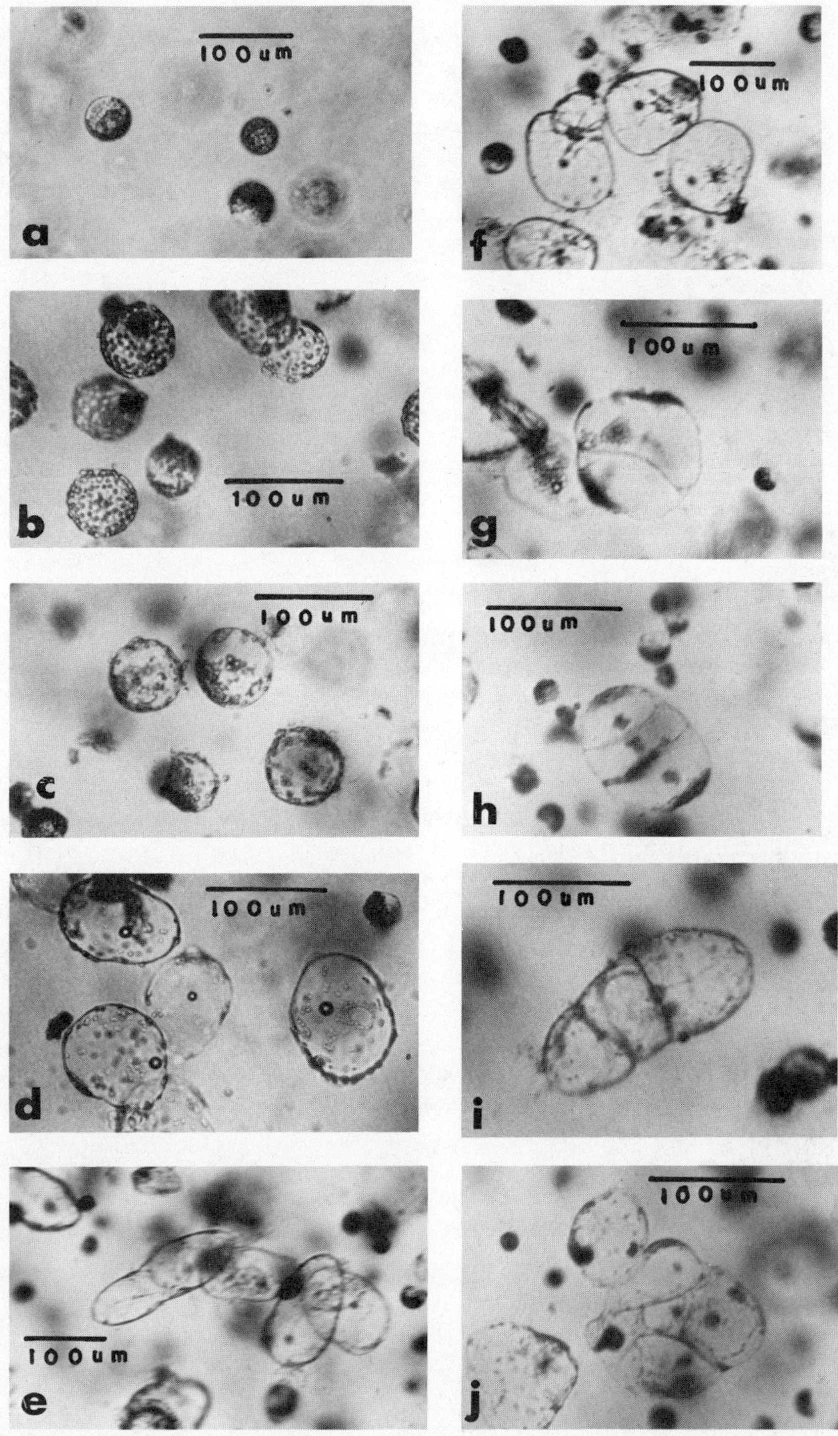
100 um
a
100 um
b
100 um
c
100 um
d
100 um
e
100 um
f
100 um
g
100 um
h
100 um
i
100 um
j

in darkness provided by aluminum foil. Cell wall synthesis was determined by the maintenance of structure of the protoplasts under hypotonic and hypertonic conditions, by staining the walls with $KI\text{-}I_2\text{-}H_2SO_4$ solution (Johansen, 1940), and by the fluorescence of Calcofluor white.

Results

Protoplasts in liquid droplet culture in petri dishes enlarged, but the cytoplasm ultimately clumped and development ceased. In agar, however, about 60% of the protoplasts enlarged and became oval-shaped, while those protoplasts which did not develop remained small and dense. With further development, the enlarged cells assumed irregular shapes (Fig. 1) and nuclei and cytoplasmic strands and cytoplasmic streaming became visible. The cultures were maintained in this manner for about six weeks, but under no conditions of light intensity or duration was a cell division observed.

In a comprehensive series of later experiments, we changed concentration of 2, 4-D and cytokinins, as well as the ratio of these compounds; we added coconut milk, substituted sorbitol for mannitol in whole or in part and varied the Ca^{++} concentration (Brenneman and Galston, in press). None of these procedures helped. However, introduction of biotin (0.08 mg/l) into the medium induced division, and "doubles", "triples" and "multiples" were frequently observed (Fig. 1). Unfortunately, such "microcalluses" did not develop further. Nuclear division and cytokinesis were unambiguously shown by acetocarmine (Fig. 2) and calcofluor staining (Fig. 3), respectively.

Substitution of 1% cellulysin for the Onozuka enzyme permitted protoplasts to be obtained within 1-2 hours at 24° C. At a concentration of about 8×10^5 per 4 ml medium, such protoplasts produced "doubles.. with great frequency but did not develop significantly beyond this point.

Constabel and co-workers (1973) placed pea leaves in the dark for 30 hours prior to isolating protoplasts. This prompted us to investigate the effect of light and dark adaptation of the plants on protoplasts isolated from leaves. Plants were grown for six days on a 16 hour photoperiod and were then placed in the dark for 16 hours prior to protoplast preparation. The entire experiment was performed under a dim green safelight. Duplicate samples of protoplasts (9 x 10 per petri dish) were incubated in the light at 400 lux and 200 lux, and in the dark. The plates in the dark were examined only with green light in a dark room and were never exposed to light.

Protoplasts grown at 200 lux were larger, reaching sizes up to 90 um x 800 um, whereas those at 400 lux were no larger than 90 um x 400 um. The cells in the dark developed more slowly and were not as elongate as those in the light,

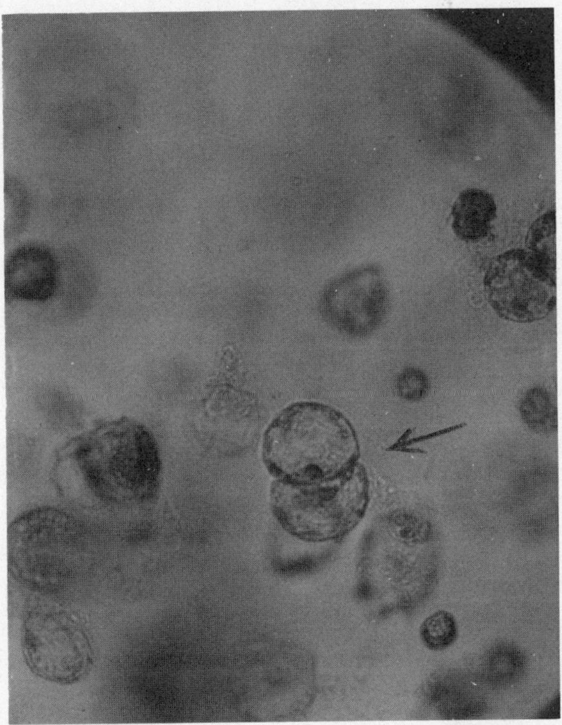

Fig. 2. Acetocarmine staining of nuclei in oat cells derived from a single protoplast. Technique according to W. A. Jensen, Botanical Histochemistry, Freemen & Co., San Francisco, 1962.

but they were more nearly symmetrical (200 um x 220 um). In all cases, doubles and triples were observed. Protoplasts from dark-adapted plants developed in almost the same way whether in the dark or in the light.

Protoplasts were also isolated from light adapted plants and were incubated in the light or dark. In these protoplasts too, development was slower in the dark. Many cells grown at 200 lux were larger than those at 400 lux (135 um x 540 um to 135 um x 360 um), whereas cells in the dark were more symmetrical. Deterioration followed the same pattern.

It seems appropriate to conclude that oat protoplasts can be readily isolated in large numbers from leaves of seedlings and can be cultured in a medium which allows cell wall formation and some cell division to occur. After 5-6 weeks in culture, however, the cells deteriorate.

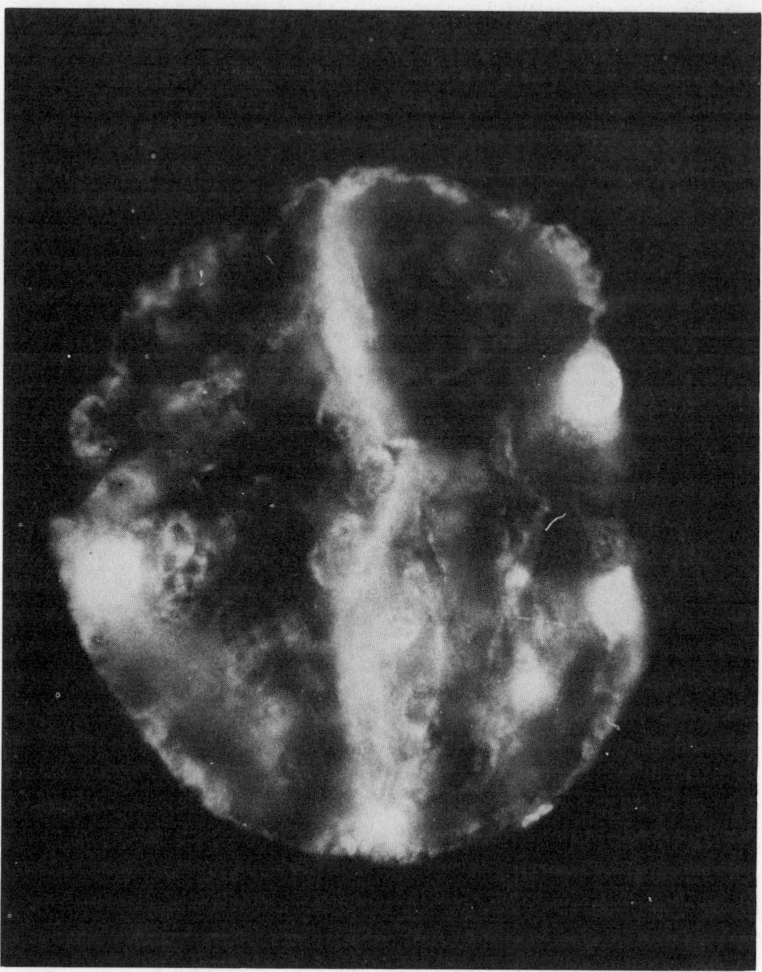

Fig. 3. New wall formation as shown by Calcofluor white staining. The pigment absorbs at 350-370 nm and emits at 440 nm. Examined in a Leitz Ortholux microscope, illuminated by a mercury lamp with appropriate suppression filters.

CULTURE OF OAT CALLUS TISSUES

Initiation of callus

Seeds of seven varieties of *Avena sativa* L. were husked and sterilized for 15 minutes in Clorox: H_2O (1:2), and for one minute in 70% ethanol. They were washed several times with sterile water, blotted, and placed at 24° C in 18 hours of light per day on LS (Linsmaier and Skoog, 1965) medium modified to contain 5 mg per liter 2, 4-D. This modified medium was called MLS. The seeds germinated and produced shoots, roots and root hairs, and some undifferentiated tissue at the hypocotyl region. This tissue when explanted did not grow well. However, when the mass of root and undifferentiated tissue was macerated and placed on medium, callus tissue did grow, as did pieces of the original plant tissue forming plantlets. Further subculture, therefore, involved a separation of callus tissue from this pre-formed and growing plant tissue.

The callus did not grow well on the MLS medium, but small green nodules appeared on all variations of medium tested. Callus was therefore grown on modifications of Prairie Regional Laboratory (PRL) medium. Increasing concentrations of cytokinins inhibited growth, occasionally killed callus, and enhanced the formation of nodules with root hairs. On PRL medium lacking cytokinins and containing half normal salt concentration and 5 mg/l 2, 4-D, callus growth markedly improved, nodule formation diminished, and good pure callus growth was obtained. Yeast extract, casein hydrolysate, corn milk and coconut milk either inhibited growth or killed the callus.

Differentiation of callus

Callus from the hypocotyl region of germinating seedlings which had been subcultured at least six times was placed on PRL/2 medium containing different concentrations and ratios of growth factors and cytokinins. When 2, 4-D was 2 mg/l or less (or preferably absent), roots with root hairs were formed (Fig. 4). When 2, 4-D was at 5 mg/l, increasing cytokinin levels permitted roots with hairs to form. With kinetin in the medium, the callus grew slowly, but with benzylamino purine (BAP), 6 (γ, γ-dimethylallylamino) - purine (2 iP), or zeatin, no callus growth occurred. With IAA, NAA and gibberellic acid, roots were formed at all concentrations tested. It is noteworthy that roots grew longer on the medium containing gibberellic acid. With cytokinins alone (0.1 - 5.0 mg/1 BAP, IP or kinetin), in the absence of 2, 4-D, roots formed and the callus tissue died.

While roots formed under all conditions tested, shoots never formed. Some stem-like organization without root formation occurred on MLS medium. Elongated (2-3 mm long), bright green nodules appeared, independent of the concentrations of growth factors tested. On a variation of the MLS medium containing major salts ½ x, minor sales ½ x, iron ½x, phosphate 1 x and organic components ½ x (called MLS var.), these bright green elongate nodules formed in especially large numbers. If 2, 4-D was omitted from the medium, roots formed.

All tissues were fixed in Farmers Acetic Acid, stained with safranin and fast green, and sectioned. Callus, subcultured on B/2 medium at least six times, had meristematically active cells throughout (Fig. 4). There were rare tracheids.

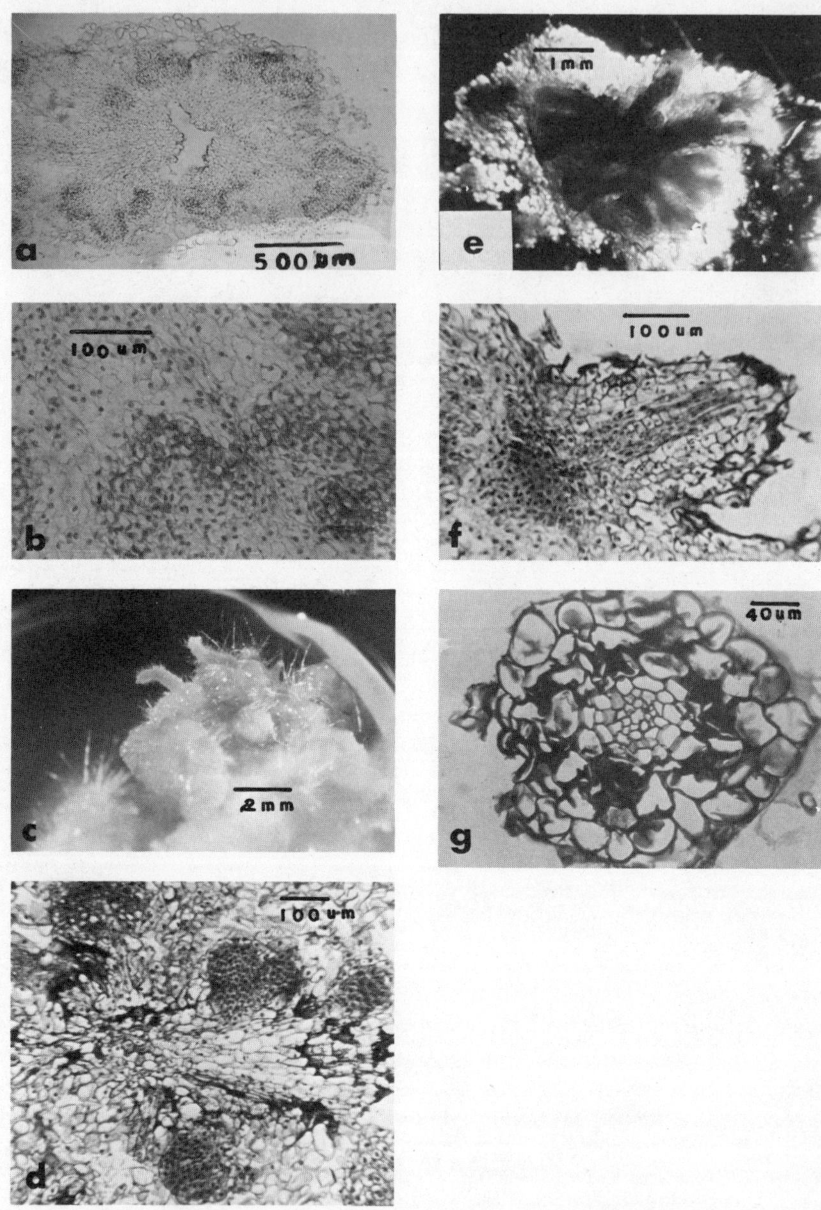
a
500 µm
b
100 um
c
2 mm
d
100 um
e
1 mm
f
100 um
g
40 um

A young nodule (about 2 mm long) had lobes of meristematically active cells, arranged along an axis of elongated cells including tracheids. When a nodule continued to develop, a many armed, red and green pigmented, rigid body frequently formed. The arms contained elongated cells and many tracheids, as well as meristematically active cells at the base of the arms. In cross section, tracheids were visible, but there was no orderly arrangement of vascular tissue.

MACROMOLECULAR SYNTHESIS IN OAT PROTOPLASTS

Protoplast preparation

About 4 grams of leaves were stripped of their lower epidermis and were floated for two hours at 32-33° C, with their lower, peeled surface in contact with 30 ml of a 5% cellulysin solution in 0.6 M mannitol at pH 5.7 (pH was adjusted with 0.1 N NaOH) in a 15 cm diameter petri dish. The released protoplasts were collected by centrifugation at 500 rpm for 15 minutes in a clinical centrifuge and washed twice with 0.6 M mannitol (pH 5.7, without adjustment). Unless otherwise specified, the protoplasts were suspended in 0.6 M mannitol at the end of the washing procedure, and counted in a hemacytometer. One ml suspensions containing about 10^6 protoplasts per milliliter were used in all experiments.

Incorporation studies

The following labelled materials were employed: L-leucine 4,5-^{3}H(N), specific activity 30.7 Ci/m mole (New England Nuclear Co.); uridine 5-^{3}H, specific activity 8.0 Ci/m mole (Schwartz/Mann); and thymidine methyl-^{3}H, specific activity 20.0 Ci/m mole (New England Nuclear Co.).

All incubations were performed in duplicate or triplicate in 10 ml Erlenmeyer flasks containing 1.0 ml protoplast suspension, shaken at 40 reciprocal strokes per minute in a Dubnoff metabolic shaking incubator at 23° C. Incorporation was studied by the methods of Mans and Novelli (1961) and Byfield

Fig. 4. Oat callus.

- *a.* *Internal distribution of meristematically active cells within callus grown on B/2.*
- *b.* *Enlarged portion of (a).*
- *c.* *Regeneration of roots and root hairs in callus on auxin-free medium.*
- *d.* *Section through callus with nodules showing distribution of elongated cells and lobes of meristematically active cells.*
- *e.* *Many armed starfish-shaped body which developed on modified Linsmaier-Skoog medium (with 5 mg/l 2, 4-D).*
- *f.* *Internal morphology of (e) showing meristematically active cells at base of protuberances containing elongated cells.*
- *g.* *Cross section of root formed on medium containing gibberellic acid (without 2, 4-D).*

and Scherbaum (1966) in which trichloroacetic acid (TCA) precipitation of macromolecules on filter-paper discs was employed. Discs of Whatman No. 3 MM chromatography paper, 2.4 cm in diameter, were numbered and suspended on straight pins stuck into a mounting board. Aliquots (100 ul) of protoplast suspension, removed at intervals during the course of the experiment, were pipetted onto the discs, which were then removed with a forceps and dropped directly into a large volume of ice-cold TCA (5% w/v for nucleic acids and 10% w/v for protein) contained in a single Erlenmeyer flask. At the end of each experiment, the accumulated discs were washed by pouring off the original TCA, adding a new charge of 5 ml of 5% TCA per disc to the original flask, swirling for 15 minutes, and repeating three times. Residual water and TCA were then removed by washing the discs twice in ether:ethanol (1:1). This step, in which most of the chlorophyll was washed out of the discs, was followed by two rinses in anhydrous ether (5 ml/disc). Each of these latter manipulations lasted about 5 minutes. At this point, the discs were dried and prepared for counting by placement on large sheets of filter paper, removing the pins and allowing them to stand in rapidly moving air for 10 minutes. Counting was performed by transferring each disc to a scintillation vial, adding 10 ml of Bray's solution (Bray, 1960), and loading into an Ansitron scintillation counter.

RESULTS

Incorporation of L-leucine

About 20% more leucine was incorporated into TCA-insoluble material in 0.6 M mannitol than in Gamborg's B5 medium. Replacing part of the mannitol with glucose or inositol reduced leucine incorporation. Addition of 10 mM $CaCl_2$ and 1 mM $MgSO_4$ to the mannitol incubation medium yielded no beneficial effects.

Protoplasts incubated with labelled leucine (1 uCi/ml) showed a linear increase in total incorporation with time for 6 hours (Fig. 5). Cycloheximide (20 ug/ml) inhibited ^{3}H-leucine incorporation by 68% after 2 hours of incubation, while actinomycin D (20 ug/ml) was without effect.

A mixture of penicillin and streptomycin (GIBCO, 10,000 units penicillin and 10,000 ug streptomycin per ml), was added at different dilutions in certain experiments to control bacterial contamination in overnight incubations. Microscopic examination showed that 100 ug/ml streptomycin was effective, but also inhibited leucine incorporation. The greater the concentration of the antibiotic, the greater was the percent inhibition (Fig. 6) which was already obvious after two hours. After 22 hours, the extent of inhibition was greater, but the pattern remained the same. We assume this streptomycin effect was due to attachment to ribosomes and concomitant reduction of protein synthesis, since poly-U and

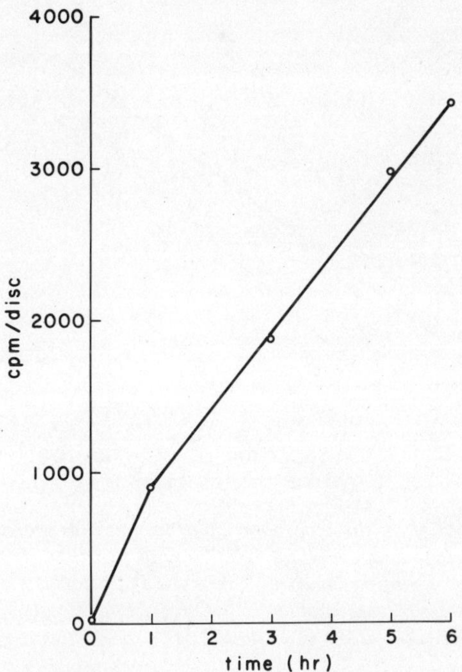

Fig. 5. Incorporation of ^{3}H-leucine into TCA insoluble material. Nondividing, six day old, oat leaf protoplasts (approximately 10^{6} in 1 ml in a 10 ml Erlenmeyer flask) were incubated with 1 uCi leucine. Aliquots of 100 were sampled for counting.

similar materials added to the external medium abolished the inhibitory effect. The addition of various amounts of unlabelled, carrier L-leucine to a constant level of ^{3}H-leucine per ml showed that the greater the leucine level, the greater was the incorporation into protein (Fig. 7).

The protoplasts lose their ability to incorporate L-leucine 6 to 9 hours after their preparation and washing. When leucine was fed immediately after washing, incorporation took place at a steady rate for 6 hours and then practically stopped. However, if L-leucine was added 2 or 6 hours after washing, similar rates of incorporation persisted for only 5 and 3 hours, respectively, before stopping (Fig. 8). There was no additional L-leucine incorporation after 10 hours of

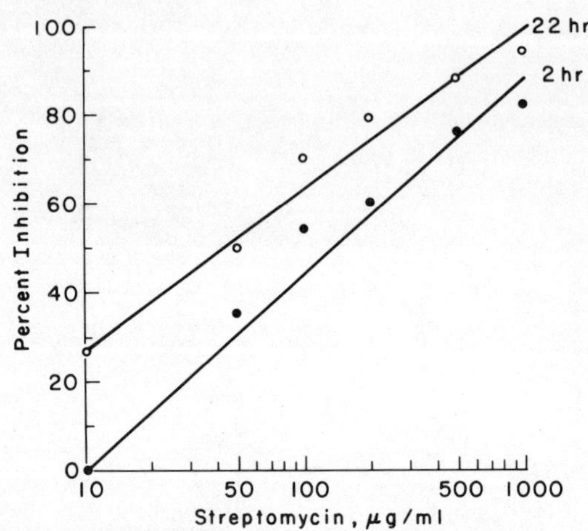

Fig. 6. Effect of antibiotics on the incorporation of leucine into proteins. Various dilutions of GIBCO penicillin-streptomycin mixture (10,000 units penicillin, 10,000 ug streptomycin per ml) were added to 1 ml of an incubation mixture of protoplasts and L-leucine (1 uCi/ml). The reaction flasks were incubated at 23° C for 22 hours, and samples for incorporation determinations were taken at 2 and 22 hours.

incubation, irrespective of the time of leucine addition to the protoplasts. The number of intact protoplasts declined linearly during the incubation, from 1.6×10^6 per milliliter at the beginning of the experiment to about 0.6×10^6 per milliliter after 20 hours. In the presence of the antibiotic mixture, at a level containing 100 ug/ml streptomycin, similar rates of L-leucine incorporation were obtained.

Auxins (2, 4-D, IAA and NAA) added to the protoplasts at final concentrations between 10^{-9} and 10^{-4} M together with ^{3}H-L-leucine, had no effect on the rate or extent of leucine incorporation into proteins.

Kinetin (10^{-9} to 10^{-4} M) was added to the protoplasts together with ^{3}H-leucine inhibited the incorporation of ^{3}H-leucine and the higher the concentration the greater was the inhibition of leucine incorporation (Fig. 9). This paradoxical effect does not accord with expectations based on experiments with excised entire leaves (Martin and Thimann, 1972).

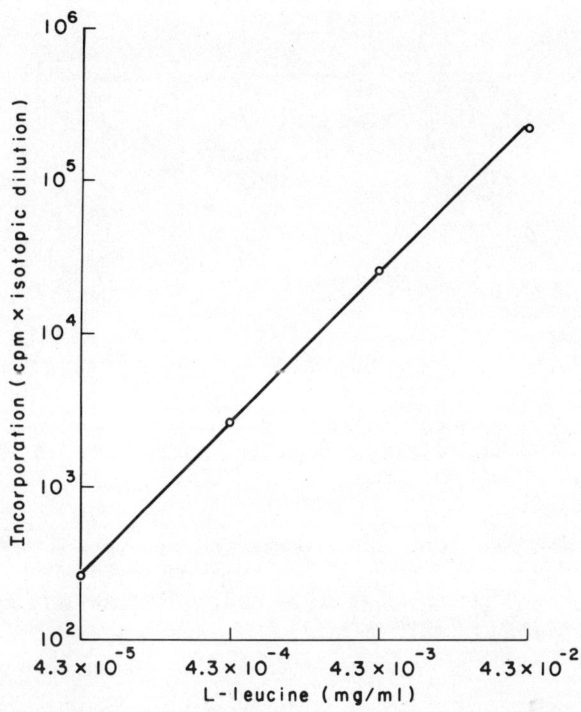

Fig. 7. Incorporation of L-leucine into protoplast proteins as a function of the concentration of leucine in the medium. Flasks with 1 ml of protoplast suspension containing 1 uCi ^{3}H-leucine per flask and various concentrations of unlabelled leucine in 0.6 M mannitol were incubated at 23° C for 2 hours.

Uridine incorporation

Incubation of ^{3}H-uridine (5 uCi/ml) with protoplasts resulted in the progressive incorporation of label into TCA insoluble matter. Using the GIBCO antibiotic mixture (penicillin-streptomycin) at a level containing 100 ug streptomycin per ml protoplast suspension in 0.6 M mannitol, the incorporation of uridine was linear for about 5 hours, but no additional net incorporation occurred during the next 16 hours (Fig. 10). Increasing uridine concentration over

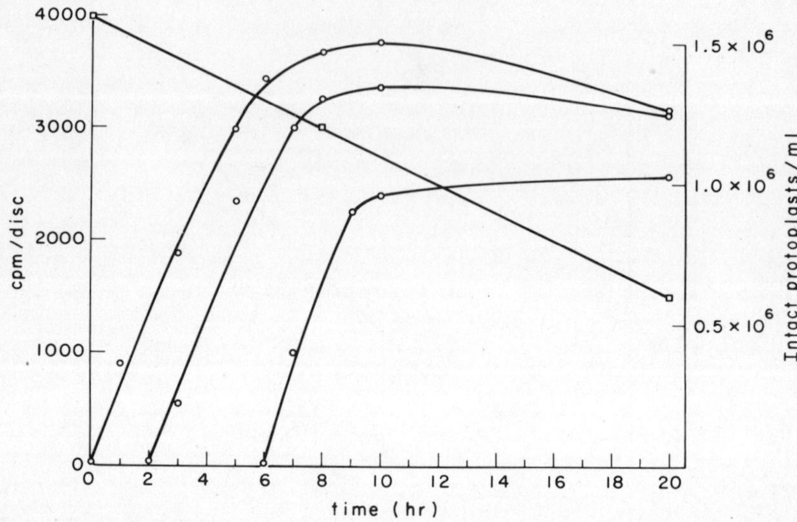

Fig. 8. L-leucine incorporation as a function of age of protoplasts after isolation. ^{3}*H-leucine (1 uCi) was added to each flask containing 1 ml protoplast suspension 0,2 or 6 hours after washing them free of enzyme. The flasks were sampled (50 ul aliquots) periodically. Protoplasts were counted with a hemacytometer.*

100 fold by adding non-labelled carrier uridine to the incubation solution, indicated that the higher the uridine concentration the greater was the incorporation.

The incorporation of uridine was not affected by 2, 4-D at concentrations between 10^{-9} and 10^{-7} M, but higher concentrations of hormone were inhibitory during a 2.5 hour incubation. Gibberellic acid and kinetin(10^{-9} to 10^{-4} M) were without effect over a 2 hour period.

Incorporation of thymidine into TCA insoluble matter

Unlike leucine and uridine incorporation, thymidine incorporation continues for at least 21 hours (Fig. 10). During the initial 5 hours the rate of thymidine incorporation is slower than that of leucine and uridine but it remains reasonably constant for the remainder of the 21 hours. Neither 2, 4-D nor kinetin (10^{-9} to 10^{-4} M) affected thymidine incorporation into protoplasts.

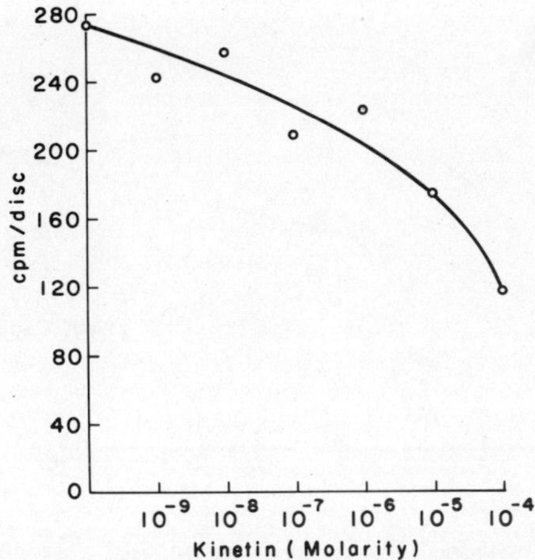

Fig. 9. Effect of kinetin on leucine incorporation into proteins. Kinetin (10^{-7} to 10^{-4} M) was added together with ^{3}H-L-leucine. Incorporation was determined after 2 hours of incubation at 23° C.

RESPONSE OF OAT PROTOPLASTS TO THE PHYTOTOXIN VICTORIN

Helminthosporium victoriae, the causative agent of Victoria blight of oats (*Avena sativa* L.), was first described thirty years ago (Meehan and Murphy, 1946). The pathogen produces a host-specific toxin (victorin) that can induce the same symptoms in susceptible cultivars as the fungus itself (Litzenberger, 1949; Luke and Wheeler, 1955; Scheffer *et al.*, 1964). The first attempt to produce the toxin *in vitro* revealed that diluted (1:45) cell-free filtrates were toxic only to susceptible tissue (Meehan and Murphy, 1947). Since this first demonstration, many attempts have been made to isolate and characterize victorin (Luke and Wheeler, 1955; Pringle and Braun, 1957, 1958, 1960; Scheffer and Pringle, 1963; Scheffer *et al.*, 1964; Dorn and Arigoni, 1972). Chromatography of concentrated cell-free filtrates on an acid alumina column resulted in a substance that was toxic to susceptible oat tissue at 2×10^{-2} g/ml (Pringle and Braun, 1957). The purified toxin is active at 2×10^{-4} g/ml (Scheffer and Pringle, 1963).

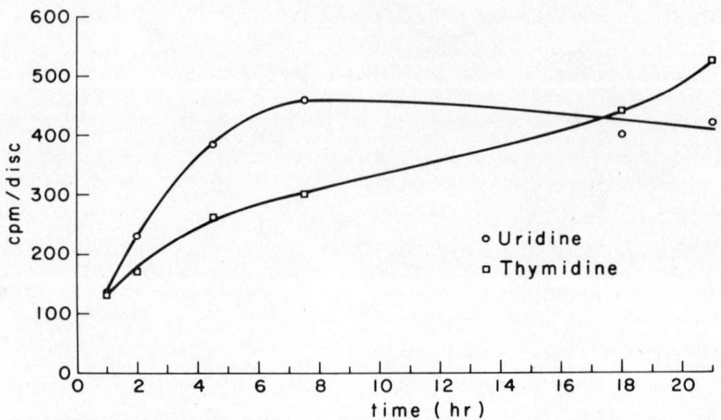

Fig. 10. Uridine and thymidine incorporation by oat leaf protoplasts. Protoplasts were incubated for 21 hours, at 23° C, with 5 uCi/ml ^{3}H-uridine or 10 uCi/ml thymidine and penicillin-streptomycin mixture (100 ug/ml streptomycin).

Although its exact structure is still unknown, the intact toxin is ninhydrin negative when first isolated, but a mild alkaline treatment results in two ninhydrin positive fractions when chromatographed on paper (Pringle and Braun, 1958). One of these spots is a tricyclic secondary amine (victoxinine, $C_{17}H_{29}NO$) and the other is a peptide containing aspartic acid, glutamic acid, glycine, valine and leucine. The nature of the bond between the base and the peptide is still undetermined (Luke and Gracen, 1972).

In susceptible oat cultivars, victorin has been shown to affect many processes, while having no significant effect on resistant tissue at low concentrations. Some of the reactions of susceptible tissue to victorin are listed below.

1. Increased 0_2 uptake (Scheffer and Pringle, 1963).
2. Loss of electrolytes (Wheeler and Black, 1963; Samaddar and Scheffer, 1968; Damann *et al.*, 1974).
3. Inhibition of phosphorus uptake and incorporation into organic compounds (Samaddar and Scheffer, 1968).
4. Inhibition of uptake and incorporation of amino acids (Samaddar and Scheffer, 1968).

5. Rapid lysis of isolated protoplasts in the presence of victorin (Samaddar and Scheffer, 1968).
6. Partial protection by carbonyl-binding reagents against toxin-induced leakage of ions (Samaddar and Scheffer, 1971).
7. Decreased toxin-induced leakage by brief pre-treatment with cycloheximide and sulfhydryl-binding compounds (Gardner and Scheffer, 1973).
8. Increased growth of coleoptile segments at high concentrations (Saftner and Evans, 1973).
9. Depolarization of membrane potentials in oat roots (Novacky and Hanchey, 1974).
10. Increased production of ethylene in detached leaves (Shain and Wheeler 1975).
11. Altered isoperoxidase patterns (Novacky and Wheeler, 1969).

Contrary to the evidence that resistant tissue is not affected by the toxin, it has been reported (Wheeler and Doupnik, 1969) that high concentrations of victorin can induce physiological changes such as the leakage of electrolytes, inhibition of root growth and an increase in respiration in resistant tissue. It is generally agreed by most researchers, however, that the toxin is highly host-specific for susceptible tissue and that the plasma membrane is the primary site of action of victorin. It is further hypothesized that the toxin binds with a specific receptor site on susceptible tissue and that the resulting membrane damage is the primary site which leads to the other effects of the toxin. Resistant tissue is thought to lack the receptor site or have sites with lower affinity for the toxin (Scheffer and Yoder, 1972). Workers agree that a single dominant gene controls the susceptibility of oats to *Helminthosporium victoriae* (Litzenberger, 1949; Finkner, 1953; Rivers, 1959).

A crude sample of victorin was supplied by M. L. Evans and R. A. Saftner, Ohio State University. The crude sample inhibited the growth of susceptible oat roots when diluted $1{:}10^4$, while it had no effect on the growth of resistant roots at a dilution of 1:10.

Effect of victorin on protoplasts

Victorin caused rapid bursting of protoplasts extracted from leaves of a susceptible variety (Park) but not from a resistant variety (Victory)(Figure 11). A dilution of 1:1000 was effective against Park protoplasts, causing 50% bursting within 1 hour, and 100% bursting after overnight incubation. By contrast, even 1:10 victorin had no effect on Victory protoplasts after overnight incubation, and may even somewhat protect the protoplasts from bursting.

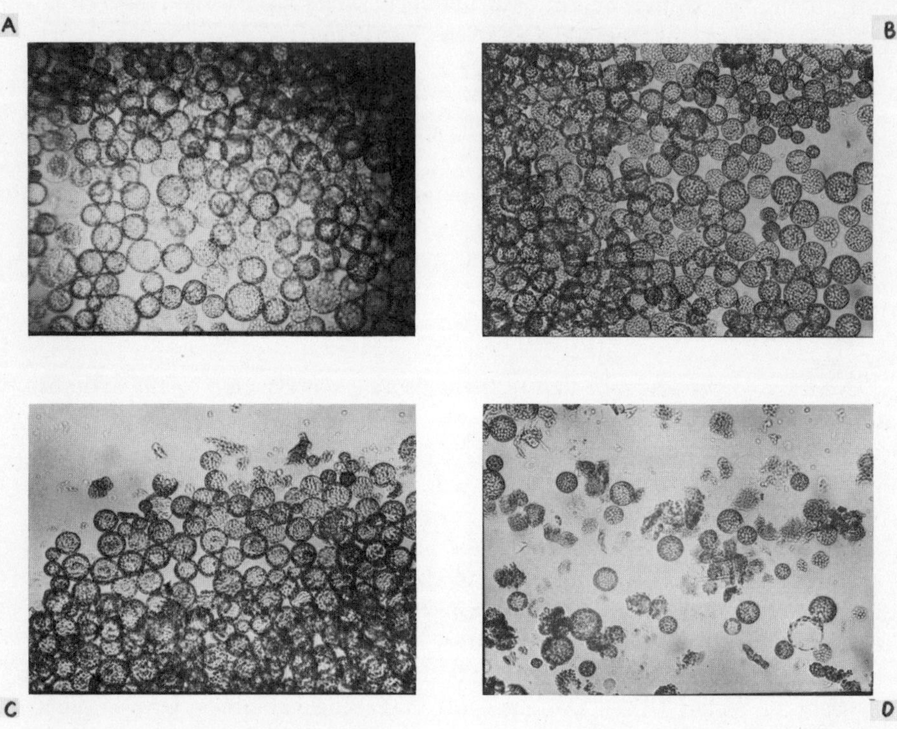

Fig. 11. Response of resistant Victory and susceptible Park oat protoplasts to identical concentration of victorin. (1 hour incubation) A. Victory, control. B. Victory, plus toxin. C. Park, control. D. Park, plus toxin.

Effect of victorin on incorporation of labelled precursors into macromolecules

Victorin greatly inhibited incorporation of leucine into proteins and uridine into RNA of sensitive Park protoplasts, but was much less effective against Victory protoplasts. Thymidine incorporation into DNA over 1-2 hour test periods was virtually unaffected by victorin in both varieties; overnight there was a 20-25% inhibition in both varieties (Tables 1 and 2).

Effect of cycloheximide

Since the most satisfactory hypothesis for toxin sensitivity is that the sensitive variety contains membrane-localized protein receptors for the toxin, we

TABLE 1
Incorporation of thymidine, uridine and leucine into macromolecules of oat leaf protoplasts.

	^{3}H-Thymidine			^{3}H-Uridine			^{3}H-L-Leucine		
	c.p.m. per million of protoplasts after								
	hrs.			hrs.			hrs.		
Protoplast type	0.5	1	2	0.5	1	2	0.5	1	2
Susceptible (cv. Park)	126	179	183	-	114	164	214	307	591
Resistant (cv. Victory)	113	164	215	-	130	168	283	371	692

studied the effect of prior incubation in cycloheximide (CHI) on the sensitivity of the oat protoplasts to toxin. Six or seven-day-old leaves of susceptible and resistant oats were peeled and floated on distilled water, either without or with 1-5 ug CHI/ml, for approximately 18 hours at room temperature in the dark. After the CHI was removed, the leaves were rinsed 3 times with distilled water and floated on the enzyme solution for protoplast preparation. The washed protoplasts were used for studies of leucine incorporation and toxin effects.

Surprisingly, leaves of both the susceptible and resistant varieties, when pre-treated with CHI, released about 2.5 times as many protoplasts as the controls (Table 3). CHI pretreatment also increased the incorporation of leucine into macromolecules of resistant protoplasts and, to a much lesser degree, of susceptible protoplasts. When a CHI concentration of 5 ug/ml was used, the increase after 0.5 and 1 hour was small, but after 2.5 and 4 hours, incorporation of leucine in resistant oats increased markedly (Fig. 12). With a concentration of 1 ug CHI per ml, the increase in leucine incorporation was also noticeable in susceptible protoplasts. By contrast, when protoplasts obtained from leaves not pretreated with CHI were exposed to leucine with or without CHI, the CHI decreased the incorporation of leucine by 70% in both types of protoplasts. These results indicate that CHI inhibited the formation of proteins which interfere with cellulolytic release of protoplasts and of proteins which favor senescent development, as suggested by Martin and Thimann (1972).

TABLE 2
Effect of victorin on incorporation of thymidine, uridine and leucine into macromolecules of oat leaf protoplasts.

Protoplast type	Toxin dilutions	^{3}H-Thymidine			^{3}H-Uridine			^{3}H-L-Leucine		
		% decrease of incorporation relative to control after								
		hrs.			hrs.			hrs.		
		0.5	1	2	0.5	1	2	0.5	1	2
Susceptible (cv. Park)	$1{:}10^4$	-	12	10	-	12	17	12	16	40
	$1{:}10^3$	-	10	17	-	9	16	26	40	64
	$1{:}10^2$	-	7	23	-	22	34	34	53	69
Resistant (cv. Victory)	$1{:}10^3$	-	-	-	-	-	-	0	13	+7
	$1{:}10^2$	-	9	18	-	11	11	17	16	16

TABLE 3
Effect of cycloheximide on release of protoplasts from oat leaves.

Protoplast type	Millions of protoplasts /ml. from leaves pretreated in: water	cycloheximide (μg/ml)
Susceptible (cv. Park)	0.7	1.8
Resistant (cv. Victory)	0.7	1.5

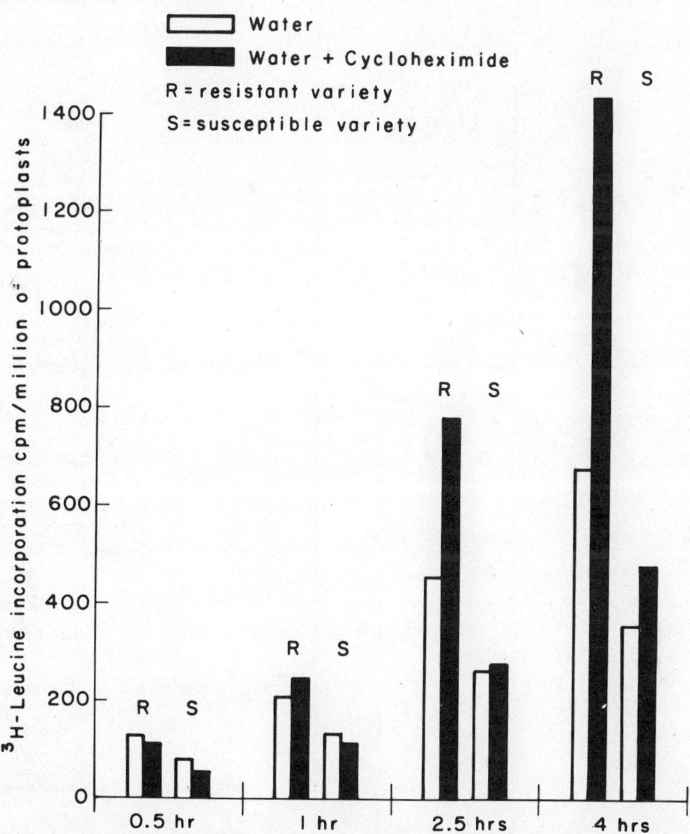

Fig. 12. Effect of cycloheximide preincubation of leaves on susceptibility of oat protoplasts to victorin, as shown by incorporation of ^{3}H--leucine. R = resistant Victory oats; 2 = susceptible Park oats.

TABLE 4
Effects of cycloheximide and victorin on incorporation of ^{3}H-L-leucine into oat protoplasts.

Protoplast type	Toxin dilution	Change in leucine incorporation (% of control) in leaves pretreated in:							
		water hours				cycloheximide (1 ug/ml) hours			
		0.5	1.0	2.5	4.0	0.5	1.0	2.5	4.0
Susceptibile (cv. Park)	$1{:}10^4$	+5	20	43	57	9	9	8	25
	$1{:}10^3$	0	21	37	53	2	12	6	11
	$1{:}10^2$	36	55	69	82	6	16	23	34
Resistant (cv. Victory)	$1{:}10^3$	6	5	9	2	4	+1	4	1
	$1{:}10^2$	32	21	24	14	13	12	15	20

Pretreatment with CHI also lowered the sensitivity of susceptible protoplasts to toxin. While toxin greatly inhibited leucine incorporation in susceptible protoplasts obtained from leaves floated on water, the inhibition was greatly reduced by pretreating leaves for 18 hours on CHI (Table 4). These results are in agreement with those of Gardner and Scheffer (1973) who reported that pretreatment of leaf segments with CHI greatly reduced toxin-induced loss of electrolytes and support the hypothesis that victorin interacts with specific protein receptors in the plasma membrane.

DISCUSSION AND CONCLUSIONS

It is clear that oat leaf protoplasts, as usually isolated by enzymatic techniques, are in a partially senescent condition. Pretreatment of the leaf with cycloheximide prevents this senescence, as manifested by increased yield of intact protoplasts and increased rates of leucine incorporation in the isolated protoplasts. Could this cycloheximide pretreatment aid in the prolongation of cell division activity in isolated protoplasts and the development of truly multicellular calluses? Experiments now under way could answer this question.

Even with calluses formed from protoplasts, we still will have to face the problems of how to induce organ differentiation. Here, there are no clear leads, but the transfer of tissues from hormone-supplemented to minimal media seems the most promising technique at present. Despite the lack of clear leads in this difficult field, these experiments must continue if the promise of protoplast technology is to be realized in cereals.

ACKNOWLEDGEMENTS

This research was aided by a grant to the senior author from the RANN program of the National Science Foundation. We thank Constance Lehman for excellent technical assistance and Drs. R. Saftner and M. Evans of Ohio State University for gifts of victorin.

REFERENCES

Aynsley, J. S. (1974). The ability of callus cultures of *Asparagus officinalis* L. to initiate roots. In *Abstracts of the 3rd International Congress of Plant Tissue and Cell Culture, #143,* University of Leicester, England.

Barba, R., and Nickell, L. G. (1969). Nutrition and organ differentiation in tissue cultures of sugarcane, a monocotyledon. *Planta 89,* 299-302.

Bayley, J. M., King, J., and Gamborg, O. L. (1972). The effect of the source of inorganic nitrogen on growth and enzymes of nitrogen assimilation in soybean and wheat cells in suspension cultures. *Planta 105,* 15-24.

Blake, J., Eeuwens, C. J., and Apavatjrut, P. (1974). Culture of coconut tissue. In *Abstract of the 3rd International Congress of Plant Tissue and Cell Culture, #223,* University of Leicester, England.

Bouharmont, J., Gonzalez-Medina, M., and Delannay, X. (1974). Production of tissues and plantlets from anther cultures of barley. In *Abstracts of the 3rd International Congress of Plant Tissue and Cell Culture, #191,* University of Leicester, England.

Bray, G. A. (1960). A simple efficient liquid scintillator for counting aqueous solutions in a liquid scintillation counter. *Anal. Biochem. 1,* 279-285.

Brenneman, F., and Galston, A. W. (1975). Experiments on the cultivation of protoplasts and calli of agriculturally important plants. I. Oats (*Avena sativa* L.). *Biochem. Physiol. Pflanz. 168,* 453-471.

Bui-Dang-Ha, D., and Mackenzie, I. (1973). The division of protoplasts from *Asparagus officinalis* L. and their growth and differentiation. *Protoplasma 78,* 215-221.

Byfield, J. E., and Scherbaum, O. H. (1966). A rapid radioassay technique for cellular suspensions. *Anal. Biochem. 17,* 434-443.

Carew, D. P., and Schwarting, A. E. (1958). Production of rye embryo callus. *Bot. Gaz. 119,* 237-239.

Carlson, P. S., Smith, H. H., and Dearing, R. D. (1972). Parasexual interspecific plant hybridization. *Proc. Nat. Acad. Sci. (U.S.A.) 69,* 2292-2294.

Carter, O., Yamada, Y., and Takahashi, E. (1967). Tissue culture of oats. *Nature 214,* 1029-1030.

Cheng, T-Y., Smith, H. H. (1974). Contributions toward an *in vitro* genetic system for *Hordeum vulgare.* In *Abstracts of the 3rd International Congress of Plant Tissue and Cell Culture, #68,* University of Leicester, England.

Churchill, M. -E., Ball, E. A., and Arditti, J. (1973). Tissue culture of orchids. I. Methods for leaf tips. *New Phytol. 72,* 161-166.

Cocking, E. C. (1960). A method for the isolation of plant protoplasts and vacuoles. *Nature 187,* 962-963.

Constabel, F., and Kao, K. N. (1974). Agglutination and fusion of plant protoplasts by polyethylene glycol. *Can. J. Bot. 52,* 1603-1606.

Cummings, D. P., C. E. Green, and D. D. Stuthman. 1976. Callus induction and plant regeneration in *Avena sativa* L. Crop Sci. 16, 465-470.

Damann, K. E., Gardner, J. M., and Scheffer, R. P. (1974). An assay for *Helminthosporium victoriae* toxin based on induced leakage of electrolytes from oat tissue. *Phyto. 64,* 652-654.

Dorn, F., and Arigoni, D. (1972). Structure of victoxinine. *J. C. S. Chem. Comm.* pp. 1342.

Doy, C. H., Gresshof, P. M., and Rolfe, B. G. (1973). Biological and molecular evidence for the transgenosis of genes from bacteria to plant cells. *Proc. Nat. Acad. Sci. (U.S.A.) 70,* 723-726.

Durand, J., Potrykus, I., and Donn, G. (1973). Plantes issues de protoplastes de Petunia. *Z. Pflanzenphysiol. 69,* 26-34.

Evans, P. K., Keates, A. G., and Cocking, E. C. (1972). Isolation of protoplasts from cereal leaves. *Planta 104,* 178-181.

Finkner, V. C. (1953). Inheritance of susceptibility to *Helminthosporium* victoriae in crosses involving *Victoria* and other crown rust resistant oat varieties. *J. Agron. 45,* 404-406.

Fodil, Y., Esnault, R., and Trapy, G. (1971). Fusion de protoplastes de coleoptiles d' Avoine. *C. R. Acad. Sci. (Paris) 273,* 727-729.

Fridborg, G. (1971). Growth and organogenesis in tissue cultures of *Allium cepa var.* proliferum. *Physiol. Plant. 25,* 436-440.

Fridborg, G., and Eriksson, T. (1974). Morphogenesis in callus cultures of *Allium cepa* var. proliferum. In *Abstracts of the 3rd International Congress of Plant Tissue and Cell Culture, #249.* University of Leicester, England.

Galston, A. W. (1948). On the physiology of root initiation in excised asparagus stem tips. *Amer. J. Bot. 35,* 281-287.

Gamborg, O. L. (1970). The effects of amino acids and ammonium on the growth of plant cells in suspension culture. *Plant Physiol. 45,* 372-375.

Gamborg, O. L., Constabel, F., and Miller, R. A. (1970). Embryogenesis and production of albino plants from cell cultures of *Bromus inermis. Planta 95,* 355-358.

Gamborg, O. L., Grambow, H. J., Kao, K. N., Ohyama, K., Steck, W., and Miller, R. A. (1972). Differentiation and hybridization of plant cells and protoplasts cultured *in vitro.* In *Proc. 4th Int. Fermentation Symposium,* pp. 677-680, Kyoto, Japan.

Gardner, J. M., and Scheffer, R. P. (1973). Effects of cycloheximide and sulfhydryl-binding compounds on sensitivity of oat tissues to *Helminthosporium victoriae* toxin. *Physiol. Plant Path. 3,* 147-157.

Giles, K. L. (1974). Complementation by protoplast fusion using mutant strains of maize. *Plant & Cell Physiol. 15,* 281-285.

Grambow, H. J., Kao, K. N., Miller, R. A., and Gamborg, O. L. (1972). Cell division and plant development from protoplasts of carrot cell suspension cultures. *Planta 103,* 347-355.

Green, C. E., and R. L. Phillips, 1975. Plant regeneration from tissue cultures of maize. *Crop Sci. 15:* 417-421.

Havranek, P., and Novak, F. J. (1973). The bud formation in the callus cultures of *Allium sativum L. Z. Pflanzenphysiol. 68,* 308-318.

Heinz, D. J., and Mee, G. W. P. (1969). Plant differentiation from callus tissue of *Saccharum* species. *Crop Sci. 9,* 346-348.

Hess, D. (1973). Transformationversuche an hohern Pflanzen: Untersuchungen zur Realisation der Exesome-Modells der Transformation bei *Petunia hybrida. Z. Pflanzenphysiol. 68,* 432-440.

Hussey, G. (1974). Tissue culture responses of bulbs and corms. In *Abstracts of the 3rd International Congress of Plant Tissue and Cell Culture, #260.* University of Leicester, England.

Iyer, R. D., and Raina, S. K. (1972). The early ontogeny of embryoids and callus from pollen and subsequent organogenesis in anther cultures of *Datura metel* and rice. *Planta 104,* 146-156.

Johansen, D. A. (1940). Plant Microtechnique. McGraw Hill, New York.

Johnson, C. B., Grierson, D., and Smith, H. (1973). The expression of plac 5 DNA within cultured cells of a higher plant. *Nature New Biology 244,* 105-107.

Jullien, M. (1973a). Division et croissance *in vitro* de cellules separees du parenchyme foliaire d' *Asparagus officinalis* L. *Z. Pflanzenphysiol. 69,* 129-141.

Jullien, M. (1973b). La culture *in vitro* de cellules separees du tissu foliaire d'*Asparagus officinalis:* de la cellule parenchymateuse au cal organogene. *C. R. Acad. Sci. (Paris), Ser. D,* 276, 733-736.

Kao, K. N., Gamborg, O. L., Michayluk, M. R., Keller, W. A., and Miller, R. A. (1973). The effects of sugars and inorganic salts on cell regeneration and sustained division in plant protoplasts. In *Protoplastes et fusion de cellules somatiques vegetales, pp. 250-255.* Coll. Int. C. N. R. S. #212.

Kawata, S., and Ishihara, A. (1968). The regeneration of rice plant *Oryza sativa* in the callus derived from the seminal root. *Proc. Jap. Acad. 44,* 549-553.

Ledoux, L., Huart, R., and Jacobs, M. (1972). Fate of exogenous DNA in *Arabidopsis thaliana.* II. Evidence for replication and preliminary results at the biological level. In *Informative Molecules in Biological Systems.* (L. Ledoux, ed.), pp. 159-175. North Holland.

Lieb, H. B., Ray, T. B., and Still, C. C. (1973). Growth of rice root-derived callus tissue in suspension culture. *Plant Physiol. 51,* 1140-1141.

Litzenberger, C. S. (1949). Nature of susceptibility to *Helminthosporium victoriae* and resistance to *Puccinia coronata* in Victoria oats. *Phyto. 39,* 300.

Loo, S. W. (1946). Further experiments on the culture of excised asparagus stem tips *in vitro. Amer. J. Bot. 33,* 156-159.

Luke, H., and Gracen, V. E. (1972). Helminthosporium Toxins. In *Microbial Toxins.* (Kadis, ed.), pp. 400. Academic Press, N.Y.

Luke, H. H., and Wheeler, H. E. (1955). Toxin production by *Helminthosporium victoriae. Phyto 45,* 453-458.

Mackenzie, I. A., Davey, M. R., and Freeman, G. G. (1974). Structure, differentiation and flavour production of onion callus isolates. In *Abstracts of the 3rd International Congress of Plant Tissue and Cell Culture,* #*255.* University of Leicester, England.

Mans, R. J., and Novelli, G. (1961). Measurement of the incorporation of radioactive amino acids in protein by a filter-paper disc method. *Arch. Bioch. Biophys. 94,* 48-53.

Maretzki, A., and Nickell, L. G. (1973). Formation of protoplasts from sugar cane cell suspensions and the regeneration of cell cultures from protoplasts. In *Protoplasts et fusion de cellules somatiques vegetales, pp. 51-63.* Coll. Int. C.N.R.S. #212.

Martin, C., and Thimann, K. V. (1972). The role of protein synthesis in the senescence of leaves. *Plant Physiol. 49,* 64-71.

Mascarenhas, A. F., Sayagaver, B. M., and Jagannathan, V. (1965). Studies on the growth of callus cultures of *Zea mays.* In *Tissue Culture* (C.V. Ramakrishnan, ed.) pp. 283-292. Dr. W. Junk, Publ., The Hague, Netherlands.

Masteller, V. J., and Holden, D. J. (1970). The growth of and organ formation from callus tissue of *Sorghum. Plant Physiol. 45,* 362-364.

Meehan, F. L., and Murphy, H. C. (1946). A new *Helminthosporium* blight of oats. *Science 104,* 413-414.

Meehan, F., and Murphy, H. C. (1947). Differential phytotoxicity of metabolic byproducts of *Helminthosporium victoriae. Science 106,* 270-271.

Melchers, G., and Labib, G. (1974). Somatic hybridization by fusion of protoplasts. I. Selection of light resistant hybrids of "haploid" light sensitive varieties of tobacco. *Molec. Gen. Genet. 135,* 277-294.

Merchant, D. J., Kahn, R. H., and Murphy, W. H., Jr. (1964), *Handbook of Cell and Organ Culture,* pp. 157-158. Burgess Pub. Co., Minneapolis, Minn.

Murakami, M., Takahashi, N., and Harada, K. (1973). Induction of haploid plant by anther culture in maize. I. On the callus formation and root differentiation. *Rep. Kyoto Prefect. Univ. Agric. 24,* 1-8.

Nickell, L. G., and Maretzki, A. (1969). Growth of suspension cultures of sugarcane cells in chemically defined media. *Physiol. Plant. 22,* 117-125.

Nickell, L. G., and Maretzki, A. (1970). The utilization of sugars and starch as carbon sources by sugarcane cell suspension cultures. *Plant & Cell Physiol. 11,* 183-185.

Niizeki, H., and Oono, K. (1968). Induction of haploid rice plants from anther culture. *Proc. Jap. Acad. 44,* 554-557.

Nishi, T., Yamada, Y., and Takahashi, E. (1968). Organ redifferentiation and plant restoration in rice callus. *Nature 219,* 508-509.

Norstog, K. S. (1956). Growth of rye grass endosperm *in vitro*. *Bot. Gaz. 117,* 253-259.

Novacky, A., and Hanchey, P. (1974). Depolarization of membrane potentials in oat roots treated with victorin. *Physiol. Plant Path. 4(2),* 161-165.

Novacky, A., and Wheeler, H. (1969). Victorin-induced changes of peroxidase isoenzymes in oats. *Phyto. 60,* 467-471.

Ohira, K., Ojima, K., and Fukiwara, A. (1973). Studies on the nutrition of rice cell culture. I. A simple, defined medium for rapid growth in suspension culture. *Plant & Cell Physiol. 14,* 1113-1121.

Ouyang, T., Hu, H., Chuang, C., and Tseng, C. (1973). Induction of pollen plants from anthers of *Triticum aestivum* L. cultured *in vitro*. *Scientia Sinica 16,* 79-95.

Power, J. B., Cummins, S. E., and Cocking, E. C. (1970). Fusion of isolated plant protoplasts. *Nature 225,* 1016-1018.

Pringle, R. G., and Braun, A. C. (1957). The isolation of the toxin of *Helminthosporium victoriae*. *Phyto. 47,* 369-371.

Pringle, R. B., and Braun, A. C. (1958). Constitution of the toxin of *Helminthosporium victoriae*. *Nature 181,* 1205-1206.

Pringle, R. B., and Braun, A. C. (1960). Isolation of victoxinine from cultures of *Helminthosporium victoriae*. *Phyto. 50,* 324-325.

Prokhorov, M. N., Chernova, L. K., and Filin-Koldakov, B. V. (1974). Growing wheat tissues in culture and the regeneration of an entire plant. *Doklady Bot. Sci. 214,* 3-5.

Rivers, G. W. (1959). Inheritance of resistance to *Helminthosporium* blight, crown rust race 216, and stem rust 7A in oats. *Agron. J. 51,* 601-603.

Ruesink, A. W. (1971). The plasma membrane of *Avena* coleoptile protoplasts. *Plant Physiol. 47,* 192-195.

Ruesink, A. W., and Thimann, K. V. (1965). Protoplasts from the *Avena* coleoptile. *Proc. Nat. Acad. Sci. (U.S.A.) 54,* 56-61.

Saftner, R. A., and Evans, M. L. (1973). Selective effects of victorin on growth and the auxin response in *Avena*. *Pl. Physiol. 53,* 382-387.

Samaddar, K. R., and Scheffer, R. P. (1968). Effect of the specific toxin in *Helminthosporium victoriae* on host cell membranes. *Plant Physiol. 43,* 21-28.

Samaddar, K. R., and Scheffer, R. P. (1971). Early effects of *Helminthosporium victoriae* toxin on plasma membranes and counteraction by chemical treatments. *Physiol. Plant Path. 1,* 319-328.

Scheffer, R. P., and Pringle, R. B. (1963). Respiratory effects of the selective toxin of *Helminthosporium victoriae*. *Phyto. 53,* 465-468.

Scheffer, R. P., and Yoder, O. C. (1972). Host-specific toxins and selective toxicity. In *Phytotoxins in Plant Diseases.* (A. Wood, ed.), Academic Press, N. Y. 530 pp.

Scheffer, R. P., Nelson, R. R., and Pringle, R. B. (1964) Toxin production and pathogenicity in *Helminthosporium victoriae*. *Phyto. 54,* 602-603.

Schenk, R. U., and Hildebrandt, A. C. (1972). Medium and techniques for induction and growth of monocotyledonous and dicotyledonous plant cell cultures. *Can. J. Bot. 50,* 199-204.

Shain, Louis, and Wheeler, H. (1975). Production of ethylene by oats resistant and susceptible to victorin. *Phyto. 65(1),* 88-89.

Sheridan, W. F. (1968). Tissue culture of the monocot *Lilium*. *Planta 82,* 189-192.

Shimada, T., Sasakuma, T., and Tsunewaki, K. (1969). *In vitro* culture of wheat tissues. I. Callus formation, organ redifferentiation and single cell culture. *Can. J. Genet. Cytol. 11,* 294-304.

Simonsen, J., and Hildebrandt, A. C. (1971). *In vitro* growth and differentiation of *Gladiolus* plants from callus cultures. *Can. J. Bot. 49,* 1817-1819.

Stadelmann, C. J., and Kinzel, H. (1972). Vital Staining of Plant Cells. In *Methods in Cell Physiology, Vol. V.* (D. M. Prescott, ed.), pp. 325-372. Academic Press, N.Y.

Straus, J. (1954). Maize endosperm tissue grown *in vitro.* II. Morphology and cytology. *Amer. J. Bot. 41*, 833-839.

Straus, J. (1960). Maize endosperm tissue grown *in vitro.* III. Development of a synthetic medium. *Amer. J. Bot. 47,* 641-647.

Taiz, L., and Jones, R. L. (1971). The isolation of barley-aleurone protoplasts. *Planta 101,* 95-100.

Takebe, I., Labib, G., and Melchers, G. (1971). Regeneration of whole plants from isolated mesophyll protoplasts of tobacco. *Naturwissen. 58,* 318-320.

Tamura, S. (1968). Shoot formation in calli originated from rice embryo. *Proc. Jap. Acad. 44,* 544-548.

Trione, E. J., Jones, L. E., and Metzger, R. J. (1968). *In vitro* culture of somatic wheat callus tissue. *Amer. J. Bot. 55,* 529-531.

Vasil, V., and Vasil, I. K. (1974). Regeneration of tobacco and petunia plants from protoplasts and culture of corn protoplasts. *In Vitro 10,* 83-96.

Wakasa, K. (1973). Isolation of protoplasts from various plant organs. *Jap. J. Genet. 48,* 279-289.

Wang, C., Chu, C., Sun, C., Wu, S., Yin, K., and Hsu, C. (1973). The androgenesis in wheat *(Triticum aestivum)* anthers cultured *in vitro. Scientia Sinica 16,* 218-222.

Warick, R. P., and Fuchs, W. H. (1968). Tissue cultures of wheat root callus *in vitro. Naturwissen. 55,* 498-499.

Wheeler, H., and Black, H. S. (1963). Effects of *Helminthosporium victoriae* and victorin upon permeability. *Amer. J. Bot. 50,* 686-693.

Wheeler, H. E., and Doupnik, B. (1969). Physiological changes in victorin-treated, resistant oat tissues. *Phyto. 59,* 1460-1463.

Wilmar, C., and Hellendoorn, M. (1968). Growth and morphogenesis of *Asparagus* cells cultured *in vitro. Nature 217,* 369-370.

Yamada, Y., Tanaka, K., and Takahashi, E. (1967). Callus induction in rice *Oryza sativa* L. *Proc. Jap. Acad. 43,* 156-160.

Yatazawa, M., Furuhashi, K., and Shimizu, M. (1967). Growth of callus tissue from rice-root *in vitro. Plant & Cell Physiol. 8,* 363-373.

Yatazawa, M., Furuhashi, K. and Lee, K. K. (1974). Amino acid nutrition of rice callus tissue. In *Abstracts of the 3rd International Congress of Plant Tissue and Cell Culture, #134.* University of Leicester, England.

Ziv, M., Halevy, A. H., and Shilo, R. (1969). Organs and plantlets regeneration of *Gladiolus* through tissue culture. *Ann. Bot. 34,* 671-676.

Novel Cellular Associations Formed *In Vitro*

Peter S. Carlson

Cell cultures permit the juxtaposition and admixture of cells of diverse origins. The spectrum of such associations formed *in vitro* is limited by the requirements of cellular physiology and not by the complex developmental and morphological processes which are naturally restrictive in the whole plant. I will discuss two instances which demonstrate the potential of *in vitro* culture to be used as a tool with which to explore novel cellular associations. In one instance, chimeral plants were induced to differentiate from mixed calluses formed by co-culturing cells of two tobacco species. In another instance, a mutally dependent association was forced between cells of a higher plant and bacterium.

CHIMERAL PLANTS

Many varieties of plants are chimeral associations of genetically dissimilar cells (Avery *et al.,* 1959; Whitehead *et al.,* 1953). Although most of these varieties originated spontaneously, chimeras have been formed experimentally as the result of grafting (Jorgensen and Crane, 1927; Neilson-Jones, 1969). Cells from both scion and stock may contribute to callus tissue which is formed at a graft union. Adventitious buds arising from this callus occasionally produce chimeral shoots. This experimental approach has limited chimeral production to combinations between species which are graft compatible and which form callus at the graft union. It is hoped that these constraints will be lifted by the use of calluses grown in culture.

An experimental system may be designed *in vitro* in which genetically distinct cells are able to proliferate contiguously and form a chimeral association. Previous work with tissue cultures has shown that a graft-like relationship can be established *in vitro* between cells of different species (Ball, 1969; 1971). Genetic markers and nutritional requirements may be exploited to identify the chimeral product of the two constituent species. By transferring these chimeral calluses to an appropriate growth medium, it should be possible to induce the development of chimeral plants.

Although the cells of a chimeral association are in contact, they remain genetically discrete and separable components of a heterogeneous tissue. This form of association is distinguished from a somatic hybrid which may result from the fusion of the cytoplasm and nuclei of somatic cells. Much research has been directed recently toward accomplishing fusion between protoplasts of oat and maize (Power *et al.,* 1970) and soybean and barley (Kao and Michayluck, 1974). This approach offers a new tool for dissecting cellular and developmental functions and may allow the construction of new plant varieties. The experimental production of chimeral plants *in vitro* could provide a complementary technique for eventually developing new varieties of vegetatively propagated species.

The *in vitro* synthesis of chimeral plants requires three sequential steps: 1) initiation, recognition and maintenance of chimeral callus; 2) regeneration of mature plants from chimeral callus; 3) verification that the recovered plants are chimera. The experimental system to be described utilizes callus cultures of *Nicotiana tabacum* cv. Wisconsin 38 (tobacco) and the amphiploid hybrid that resulted from the cross of *N. glauca x N. lansdorfii* with subsequent chromosome doubling. Throughout this study the very distinct morphologies and hormonal requirements of *N. tabacum* and the amphiploid hybrid were exploited as markers to identify the tissues of these two forms. Preliminary work was also carried out utilizing tissues from *Solanum melongena* cv. Black Beauty (egg plant); *Lycopersicon pimpinelli folium,* P. I. 212408 (tomato); and *Glycine max* cv. Kanrich (soybean).

Chimeral Callus

Chimeral callus was most easily obtained by placing pith slices of the two different genotypes adjacent to and in contact with each other on a medium which contained the mineral salts and vitamin concentrations described by Linsmaier and Skogg (1965) and 4% sucrose. The medium also contained the auxin, indoleacetic acid (IAA) at 3 ug/ml and the cytokinin, 6(γ,γ-dimethylallylamino)-purine (2iP) at 0.3 ug/ml. This medium promotes callus formation in both *N. tabacum* and the amphiploid hybrid. The cultures were maintained at 22-24° C on a 16 hour light - 8 hour dark cycle. After callus initiation and proliferation, the chimeral regions were excised from the surrounding tissue and subcultured. The distinct morphology and color of callus derived from the two species allowed the chimeral regions to be identified visually since the *N. tabacum* formed a more friable and light colored tissue than the amphiploid hybrid. With each successive transfer, the two cell types became more interdispersed.

In another approach to the construction of a chimeral callus, suspensions of protoplasts from the two species were mixed and plated together. This method was unsuccessful because the chimeral associations could not be identified positively from among the many regenerated calluses. This result may reflect the difficulty of identifying heterogeneous associations. In fact, if very distinct genetic markers are used, the plating of mixed cell suspensions may prove a more efficient method of producing chimeral calluses.

Differentiation Of Mature Plants

Mature plants of *Nicotiana tabacum* were recovered by sequential transfer of callus tissue to media containing different hormone concentrations. Higher relative levels of cytokinin were used initially to promote shoot formation and subsequent exposure to higher relative levels of auxin promotes root formation. In contrast, shoots were organized on callus derived from the amphiploid hybrid in the absence of exogenous hormones. Root formation was not accomplished with this genotype (Table 1). The difference in the exogenous hormone levels required to induce organogenesis in the two genotypes was crucial to the efficient recovery of chimeral plants.

TABLE 1
Hormone concentrations required to induce organogenesis

Species	Shoot induction (μg/ml) IAA	Shoot induction (μg/ml) 2iP	Root induction (μg/ml) IAA	Root induction (μg/ml) 2iP
N. tabacum	0.3	3.0	1.0	0
Amphiploid hybrid	0	0	N.A.*	N.A.*
S. melongena	0.3	10.0	1.0	0
L. pimpinelifolium	0.3	10.0	1.0	0

*Not accomplished

Plants differentiated from a chimeral callus may be composed of tissue from one or both of the two constituent species. The number of chimeral individuals were enriched by a phenotypic rescue technique. The chimeral callus was placed on a medium which favored organ development in tissue of one genotype and shoots displaying a morphology characteristic of the other genotype were selected.

In this study approximately 300 separate chimeral calluses were cultured for at least six months as visually identifed chimeras. They were transferred to fresh medium every three weeks. When these chimeral calluses were induced to differentiate, approximately 7000 shoots were recovered and examined. Of these shoots, 237 were selected as potential chimeras and were regenerated into mature plants. Twenty eight of the 237 plants were confirmed to be truly chimeral by procedures described in the following section.

Verification Of The Chimeral Composition Of The Plants

The most direct method of determining the chimeral nature of a plant is to demonstrate the cells of each tissue layer in the apex contain a chromosome number or morphology characteristic of a given genotype. However, chromosome numbers fluctuate greatly in cultured cells. Therefore, chromosome numbers are not reliable markers of the genotype of a tissue layer in chimeral plants obtained from *in vitro* cultures. The genotype of the tissue layers must be ascertained from other traits.

Previous research on several species of *Solanaceae* has demonstrated that the three germ layers in the apical meristem are independent and each contributes specific tissue to the developing organs. The first tunica layer (T1) gives rise to the epidermis of the stem, the second tunica layer (T2) gives rise to both the male and female gametes, and the corpus region gives rise to the central region of the pith (Avery *et al.,* 1959). Thus, the genotype of each of the three germ layers may be deduced by examining distinguishing characteristics of the derivative mature tissue.

The morphology and density of epidermal hairs on the stem are species specific characters which are reliable indicators of the composition of the T1 layer (Neilson-Jones, 1969). Since the morphology and density of epidermal hairs of *N. tabacum* and the amphiploid hybird differ (Goodspeed, 1954), the genotype of the T1 was determined by inspection of the epidermal hairs on the stem of the chimeral plant. This trait proved to be a distinct and autonomous characteristic of the epidermal tissue (i.e. T1 layer) in the chimeral plants. The hair morphology of plants regenerated from callus tissue of either *N. tabacum* or the amphiploid hybrid showed the expected species specific characteristics.

The genotype of the T2 layer was the same as that of the progeny obtained from self-fertilization of the chimeral plant. In all cases, only one parental type, either *N. tabacum* or amphiploid hybrid, was recovered among the progeny from self-fertilization of a chimeral individual. Approximately 30 progeny from each selfed presumptive chimeral plant were scored.

The genotype of the corpus region was identified by explanting pieces of the central pith into *in vitro* culture. Pith tissue of the amphiploid hybrid formed callus on medium containing 0.1 μg IAA/ml whereas pith from *N. tabacum* did not.

The composition of 25 of the 28 chimeral plants recovered are described in Table 2. Not all possible chimeral associations of the two genotypes were recovered, although we know of no reason why these types should not be expected to appear if a larger sample of chimeral plants were analyzed. Analysis of three individuals was precluded by instability in the tissue layers or failure to yield viable progeny following self-fertilization. The basis of this tissue instability is unknown, but was not due to segregation in the mericlinal chimera. Many of the chimeral plants displayed some instability in the composition of their meristematic layers over long periods of growth and upon vegetative propagation. The most frequently observed instability was the replacement of amphiploid hybrid tissue by *N. tabacum* tissue.

TABLE 2
Number of chimeral plants produced between the amphiploid hybrid (Hyb) and N. tabacum (Nt) and the genotypic composition of their tissue layers.

		Genotype of layer		
	Number of individuals	First tunica layer	Second tunica layer	Corpus
	11	Nt	Nt	Hyb
	0	Nt	Hyb	Nt
	3	Nt	Hyb	Hyb
	2	Hyb	Hyb	Nt
	1	Hyb	Nt	Hyb
	8	Hyb	Nt	Nt
	3		Undetermined	
Total	28			

A range of morphologies intermediate between *N. tabacum* and the amphiploid hybrid were observed among the chimera plants. In general, the T2 layer exerted the most distinct influence on the gross morphology of the chimeral plants. The morphology of leaves from the chimeral plants is illustrated in Fig. 1. Since plants regenerated from *in vitro* cultures often show morphological abnormalities, these photographs do not demonstrate definitively that the morphological deviations in the leaves are due to their chimeral composition. It was impossible to distinguish reliably the morphological differences attributable to the effects of *in vitro* culture.

All recovered chimeral plants formed tumorous outgrowths on the stem. Tumor formation has been shown to be an autonomous genetic characteristic of amphiploid hybrid tissue and the observation that all recovered classes of chimeral plants formed tumors demonstrated that tumor production was not confined to any single tissue layer in the mature plant (with the possible exception of the T2 layer).

Several attempts were made to generate chimeral plants between *(N. tabacum* and *L. pimpellifolium,* between *N. tabacum* and *S. melongena* and between *N. tabacum* and *G. max.* These attempts were largely unsuccessful with the exception that two chimeral plants containing tissue from both *N. tabacum* and *S. melongena* were regenerated. These plants contained a T1 layer of *S. melongena* (as determined by their epidermal hair morphology) and a T2 and corpus of *N. tabacum*. Use of genetic markers expressed in regenerated shoots may permit more efficient recovery of these chimeral types. Although the technique of *in vitro* chimeral callus construction releases the experimental synthesis of chimeral plants from the requirement of graft compatability, there are undoubtedly limits to the applicability of *in vitro* methods.

The procedures which have been outlined in this section represent an extension of earlier work in which callus formed at a graft union was used to facilitate chimera production. These methods may prove useful for constructing improved varieties of vegetatively propagated crops. Since these procedures permit the formation of chimeral plants from known parental types, it may be possible to combine the desirable characteristics of different genotypes into a single variety. Chimeras synthesized *in vitro* may also provide a useful tool for the study of plant development. The contribution of each germ layer to mature organs may be followed in chimeras constructed from genetically dissimilar tissues containing appropriate identifying genetic markers.

PLANT-BACTERIAL ASSOCIATIONS

In addition to constructing associations between plant cells of different genotypes, *in vitro* culture methods may be used to bring together cells of more distant origins. A case in which this has been accomplished in nature is the agronomically important symbiotic relationships between many plant species and bacteria capable of reducing atmospheric nitrogen (Stewart, 1966; Bond, 1968). In several laboratories, *in vitro* systems are being developed for the study of these natural associations. The reduction of acetylene to ethylene, a sensitive assay of bacterial nitrogenase activity (Hardy *et al.*, 1968), has been reported in *Rhizobium*-infected soybean cultures (Holsten *et al.*, 1971; Phillips, 1974; Child and LaRue, 1974). In one instance, the bacteria were shown to be located within the cytoplasm of the infected plant cell (Holsten *et al.*, 1971). The uptake of *Rhizobium* cell by isolated protoplasts of *Pisum sativum* has been accomplished (Davey and Cocking, 1972). Recent work has also shown that non-leguminous cell cultures may also allow *Rhizobium* to produce nitrogenase (Snowcroft and Gibson, 1975; Child, 1975).

Fig. 1. The leaf morphology of N. tabacum, amphiploid hybrid and some of the chimeral plants containing tissue of both types. The composition of the T1, T2 and corpus regions are given for each individual. The N. tabacum leaf is on the left and amphiploid hybrid leaf is on the right. The composition of the leaves from left to right is, 1) T1,T2, corpus, N. tabacum; 2) T1, amphiploid hybrid; T2, corpus, N. tabacum; 3) T1, T2, amphiploid hybrid; corpus, N. tabacum; 4) T1, amphiploid hybrid; T2, N. tabacum; corpus, amphiploid hybrid; 5) T1, T2, N. tabacum; corpus, amphiploid hybrid; 6) T1, N. tabacum; T2, corpus, amphiploid hybrid; 7) T1, T2, corpus, amphiploid hybrid.

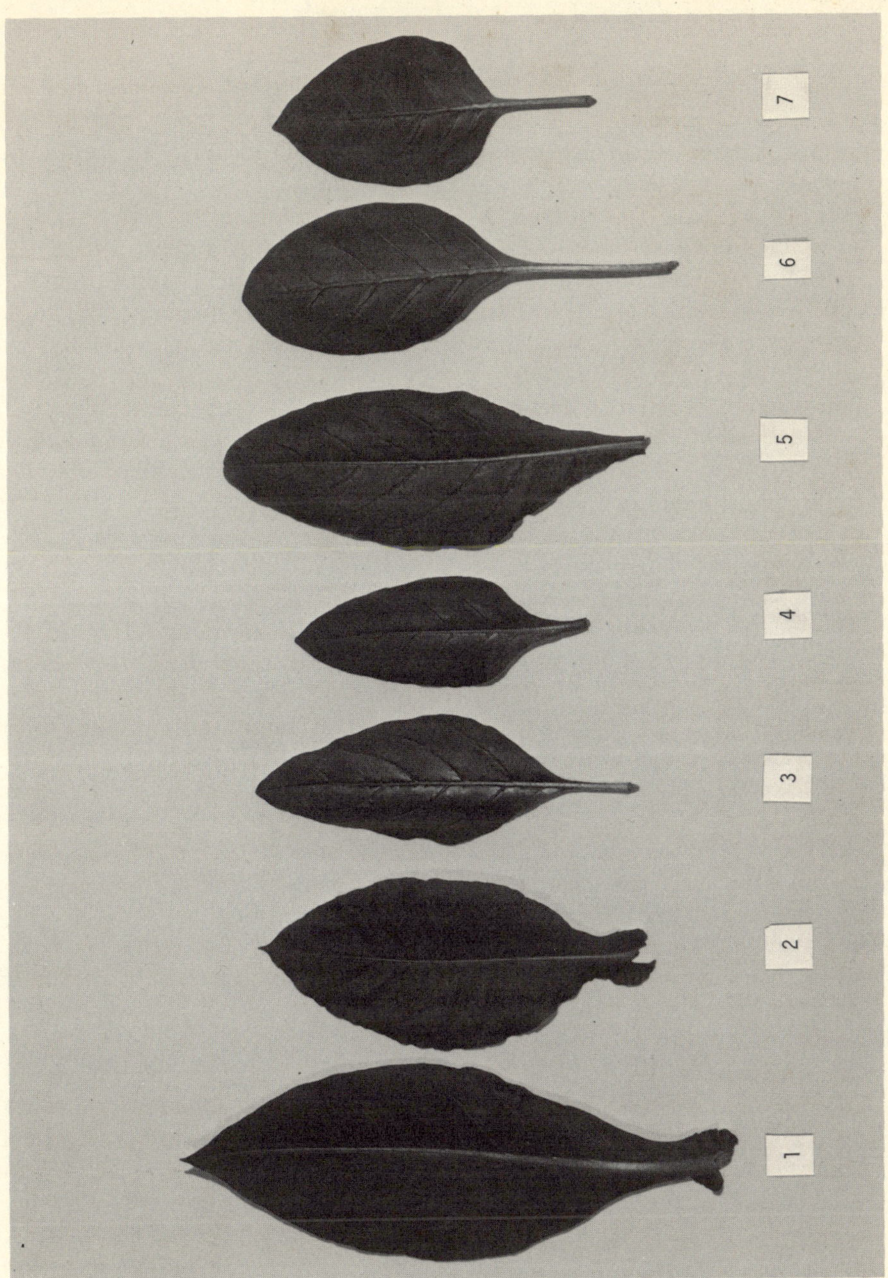
1
2
3
4
5
6
7

We have attempted to define an experimental system for investigating the possibility of extending the nitrogen-fixing associations to additional crop species. Tissue culture techniques have been used to force an association between the free-living nitrogen-fixing bacterium *Azotobacter vinelandii* and cells of carrot, *Daucus carota.* This system was based upon the establishment of a condition of mutual dependency between an auxotrophic strain of *Azotobacter* and carrot cells cultured on a medium lacking combined nitrogen. By selecting a free-living nitrogen-fixing bacterial species for these experiments, we hoped to bypass the complex interactions which have evolved in natural symbioses.

Formation of *Azotobacter*-Plant Cell Associations

Portions of a carrot cell suspension obtained from *Daucus carota cv.* Danver's Half Long (Fredonia Seed Company, Fredonia, N. Y.) were transferred to a fresh, liquid Linsmaier and Skoog (1965) medium supplemented with 4% sucrose and containing 3 μg/ml IAA and 3 μg/ml 2iP and incubated at 23° C under constant illumination. Following a three day incubation, these carrot cultures were inoculated with log phase cells of an adenine auxotrophic strain of *Azotobacter vinelandii* (American Type Culture Collection, ATCC 25308) grown on modified Burk's nitrogen-free medium (Carlson and Chaleff, 1974). The final concentration of the bacteria was 10^6 cells per ml. After 12 days incubation at 23°C under constant illumination the mixed cultures were washed and resuspended in a Linsmaier and Skoog medium from which NH_4NO_3 and KNO_3 were omitted (N-free medium). The cultures were incubated for an additional two weeks and then they were plated in N-free carrot medium solidified with 1% Noble agar. The plates were incubated at 23° C in a 16 hour light-8 hour dark cycle. The rare growing colonies were selected as they appeared three to six months after plating and were transferred to either N-free medium or a Linsmaier and Skoog medium which lacked NH_4NO_3 and contained 0.19g/l KNO_3 (Low N medium). The cultures were maintained on the same media and were transferred approximately every six weeks.

Growth Studies

Carrot cells were unable to survive *in vitro* in the absence of combined nitrogen. When 1.9 mM KNO_3 was supplied as the sole source of combined nitrogen, the carrot cell mass survived over a four month period, but did not proliferate. Calluses derived from *Azotobacter*-inoculated carrot cell suspensions increased in fresh weight on these media. This increase was significantly greater on the low nitrogen medium than on the N-free medium, indicating that nitrogen limited growth. On medium containing 20.6 mM NH_4NO_3 and 18.8 mM KNO_3 (standard concentrations) no difference in growth rates were observed between uninoculated calluses and inoculated calluses which were capable of growth on N-free medium (Fig. 2). Although electron microscopy does not reveal cell divisions in the *Azotobacter*-inoculated callus, thin cell walls were observed with sufficient frequency to suggest that the carrot cells maintained a moderate rate of mitotic activity.

Reisolation Of *Azotobacter* Cells

Azotobacter-infected calluses capable of growth on N-free or low N medium were suspended in T broth and in Burk's minimal and adenine-supplemented media. After several days, turbidity developed only in the adenine-supplemented Burk's medium. This suspension was plated and identified as the original *Azotobacter* adenine auxotroph with which the calluses had been inoculated. A three week absence of growth in T broth indicated that the calluses contain no contaminating microorganisms. T broth supports growth of a prototrophic *Azotobacter* strain but not of the adenine requiring strain. None of these three media became turbid when inoculated with control carrot callus.

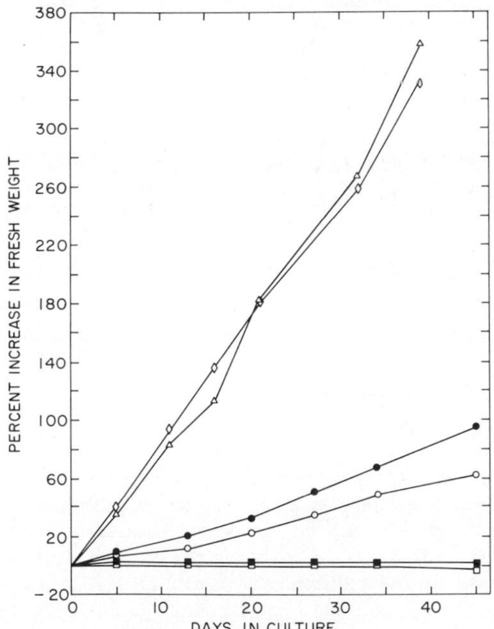

Fig. 2. Growth of Azotobacter-infected and uninfected control carrot tissues on media containing different levels of combined nitrogen. Calluses were grown on low N medium for 3 weeks before being transferred to the medium indicated. Control carrot tissue transferred to N-free (□), low N (■), or standard Linsmaier and Skoog (Δ) medium. Azotobacter containing carrot tissue transferred to N-free (o), low N (●), or standard Linsmaier and Skoog (◊) medium.

Nitrogenase Activity

Azotobacter-containing carrot tissue capable of growth on N-free medium evolved significantly more ethylene in the acetylene-reduction assay for nitrogenase than did the uninoculated control callus (Table 3). The increased ethylene production by the composite callus was observed only in the presence of acetylene and, therefore, was not due to endogenous ethylene synthesis by the plant tissue.

Penicillin Treatment

The ability of the inoculated callus to grow on N-free medium was destroyed by the addition of penicillin G at a concentration which was known to kill *Azotobacter*-cells (50 ug/ml). Growth of either inoculated or uninoculated calluses on medium containing normal levels of combined nitrogen was

TABLE 3
Acetylene-reduction activity

Tissue	Medium	Incubation Period (hr)	C_2H_2	nmol C_2H_4 ml gas
No Tissue	-	24	+	0.17
Azotobacter	N free	1	+	1.91
Carrot	+N	24	-	0
Carrot	N free	24	-	0
Carrot	+N	24	+	0.31
Carrot	N free	24	+	0.30
Carrot-*Azotobacter*	N free	24	-	0
Carrot-*Azotobacter*	N free	24	+	1.84
Carrot-*Azotobacter*	N free	0	+	0.28

Nitrogenase activity was determined by placing 50 mg of fresh carrot callus into 1 ml vials which were then injected with 0.1 ml acetylene. The *Azotobacter* controls contained approximately 8 X 10^7 cells in 0.1ml modified Burk's medium (Carlson and Chaleff, 1974). Reactions were stopped by the addition of 1.0 ml 0.1N H_2SO_4. The ethylene content of samples was determined by published gas chromatographic procedures (Burris, 1972).

not inhibited by the addition of pencillin. These observations indicated that the ability of the inoculated callus to grown on N-free medium was dependent upon the presence of functional bacterial cells in the callus mass.

Electron Microscopy

Electron micrographs of inoculated carrot callus grown on low N medium clearly established the presence of bacterial cells in the intercellular regions of the tissue (Fig. 3). *Azotobacter* cells were also found in the medium beneath the callus. No bacteria were observed within the carrot cells. Contaminating microrganisms were not detected in either the callus or the underlying medium.

The fine structure of the bacteria present in the cultured tissue (Fig. 3) were comparable in all essential details to that of *Azotobacter vinelandii* (Vela *et al.,* 1970). The bacterial cells found within the callus contained many internal

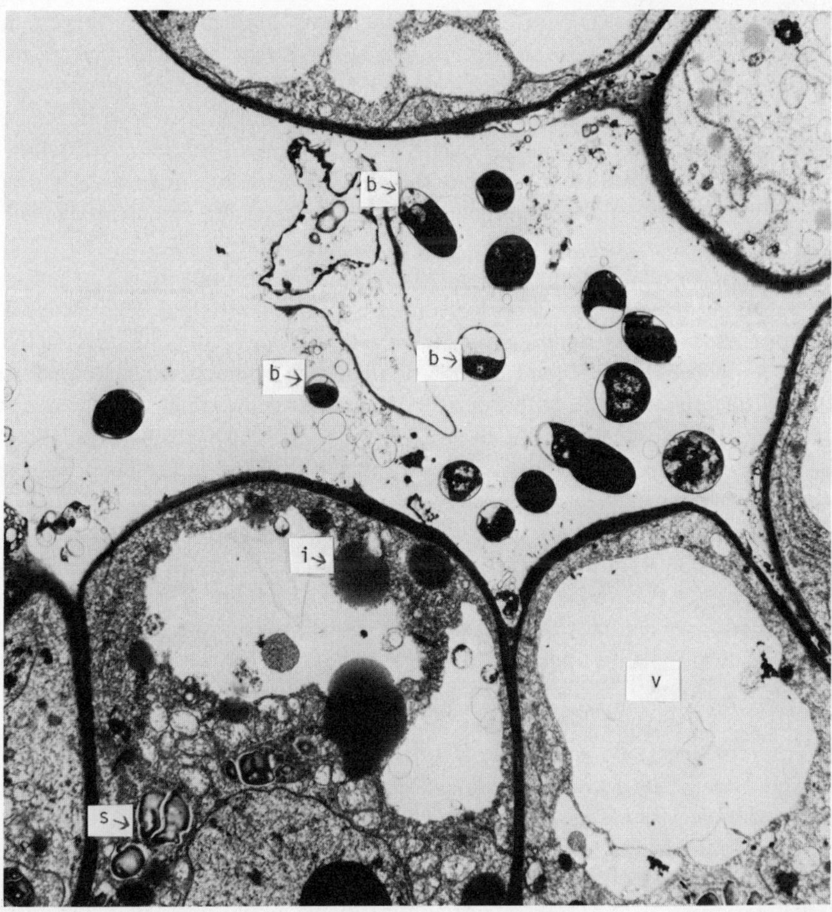

Fig. 3. Electron micrograph of a region in a carrot - Azotobacter composite callus. The carrot tissue was fixed with glutaraldehyde and OsO_4 *. The fixed samples were embedded in expoxy resin and thin sections stained with uranyl and lead salts (Carlson and Chaleff, 1974). The bacteria (b) lie outside of plant cells. The plant cells contain large vaculoes (v), electron dense inclusions (i) and starch grains (s) in addition to other normal cytological features.*

vesicles near the cell periphery. These vesicles were characteristic of *Azotobacter vinelandii* cells growing at the expense of atmospheric nitrogen. Fewer vesicles were found in bacteria utilizing NH_4 and none were visible in cells growing on NO_3 (Oppenheim and Marcus, 1970).

Carrot-*Azotobacter* Association

The independent lines of evidence which have been presented suggest that an association has been forced *in vitro* between carrot and *Azotobacter* cells. Only carrot cell cultures which have been inoculated with *Azotobacter* were able to proliferate on medium lacking combined nitrogen. These cultures have been maintained for over one year on N-free medium without any reduction in growth rate. *Azotobacter*-inoculated tissue regains its dependence on an exogenous source of combined nitrogen in the presence of penicillin. Penicillin did not affect the growth of control carrot or mixed carrot-*Azotobacter* calluses on medium containing combined nitrogen. Thus, it appears that a procaryote was providing the carrot cells with a source of reduced nitrogen. This conclusion was supported by the significantly higher levels of acetylene-reduction activity in the *Azotobacter*-inoculated calluses than in the uninfected control calluses. Finally, the presence of *Azotobacter* cells in the callus capable of growth on N-free medium was demonstrated by electron microscopy and by reisolation of the original *Azotobacter* adenine auxotroph with which the calluses had been infected.

The association between cells of carrot and *Azotobacter* was accomplished by designing a system in which the two components are able to survive only by entering into a relationship in which each complements a deficiency of the other. Presumably, carrot cells receive reduced nitrogen from *Azotobacter* cells which, in turn, depend upon the carrot cells to satisfy their auxotrophic requirement for adenine. Further experiments have attempted to better define this association. It appears that the association was not due to a genetic modification of either the plant or the bacterial cells. Both plant and bacterial cells can be recovered from the association. When these cells were used to reestablish the association, the frequency of growing calluses on N-free media was not increased over that found in the initial experiments.

The experiments reported here describe an incipient system which must be refined much further before any applications may be considered. The carrot-*Azotobacter* association was not completely stable. Occasionally, sectors of a callus lost the ability to grow on N-free medium. Other calluses became overgrown by bacteria. Because these bacteria grow in the absence of both combined nitrogen and adenine, they were prototrophic revertants of the original *Azotobacter* strain. This latter form of instability many be controlled by using nonrevertible *Azotobacter* strains in which essential genes have been deleted. The composite callus grew very slowly on either N-free or low N medium. We are attempting to increase the flow of reduced nitrogen to the plant cells by using *Azotobacter* strains which are derepressed for nitrogenase synthesis.

An assessment of the agricultural usefulness of the carrot-*Azotobacter* association must await the regeneration of mature plants from the composite calluses. Thus far, we have been unable to accomplish this. The use of genetic markers and nutritional requirements to force a symbiosis should be a generalized procedure applicable to all plant species which can be grown in culture. A novel approach to the production of useful plant-bacterial associations which may have more immediate application would attempt to modify existing marginal host-parasite relationships. The modification of endophytic bacteria to some distinctly useful function (e.g. nitrogen fixation) may produce positive agronomic effects. Considering the sophisticated state of bacterial genetics, this approach could well yield results more readily than manipulating plant cells.

ACKNOWLEDGEMENTS

I thank Dr. R. Chaleff for generously allowing me to use our joint work as a basis for this review. Drs. T. Rice, J. Polacco and D. Parke have provided thoughtful comments on this work.

REFERENCES

Avery, A. G., Satina, S., and Rietsema, J. (1959). *Blakeslee: The Genus Datura.* Ronald Press Co., New York.

Ball, E. (1969). Histology of mixed callus cultures. *Bull. Torrey Bot. Club 96,* 52-59.

Ball, E. (1971). Growth of plant tissue upon a substrate of another kind of tissue. *Z. Pflanzenphysiol. 65,*140-158.

Bond, G. (1968). Some biological aspects of nitrogen fixation. In *Recent aspects of Nitrogen Metabolism in Plants* (E. Hewitt and C. Cutting, eds.), pp. 15-25. Academic Press, New York.

Burris, R. H. (1972). Nitrogen fixation-assay methods and techniques. In *Methods In Enzymology* (A. San Pietro, ed.), pp. 415-431. Academic Press, New York.

Carlson, P. S., and Chaleff, R. S. (1974). Forced association between higher plant and bacterial cells *in vitro. Nature 252,* 393-395.

Child, J. J. (1975). Nitrogen fixation by a *Rhizobium* sp. in association with non-leguminous plant cell cultures. *Nature 253,* 350-351.

Child, J. J., and La Rue, R. A. (1974). A simple technique for the establishment of nitrogenase in soybean callus culture. *Plant Physiol., 53,* 88-90.

Davey, M. R., and Cocking, E. C. (1972). Uptake of bacteria by isolated higher plant protoplasts. *Nature 239,* 455-456.

Goodspeed, T. H. (1954), *The Genus Nicotiana.* Chronica Botanica Co., Waltham, Mass.

Hardy, R. W. F., Holsten, R. D., Jackson, E.K., and Burns, R. C. (1968). The acetylene-ethylene assay for N_2 fixation: laboratory and field evaluation. *Plant Physiol. 43,* 1185-1207.

Holsten, R. D., Burns, R. C., Hardy, R. W. F., and Hebert, R. R. (1971). Establishment of symbiosis between *Rhizobium* and plant cells *in vitro*. *Nature 232,* 173-176.

Jorgenson, C. A., and Crane, M. B. (1927). Formation and morphology of *Solanum* chimeras. *J. Genetics 18,* 247-273.

Kao, K. N., and Michayluk, M. R. (1974). A method for high-frequency intergeneric fusion of plant protoplasts. *Planta 115,* 355-367.

Linsmaier, E. M., and Skoog, F. (1965). Organic growth factor requirements of tobacco tissue cultures. *Physiol. Plant. 18,* 100-127.

Neilson-Jones, W. (1969). *Plant Chimeras.* Methuen and Co., Ltd., London.

Oppenheim, J., and Marcus, L. (1970). Correlation of ultrastructure in *Azotobacter vinelandii* with nitrogen source for growth. *J. Bacteriol. 101,* 286-291.

Phillips, D. A. (1974). Factors affecting the reduction of acetylene by *Rhizobium*-soybean cell associations *in vitro. Plant Physiol. 53,* 67-72.

Power, J. B., Cummins, S. E., and Cocking, L. C. (1970). Fusion of isolated plant protoplasts. *Nature 225,* 1016-1018.

Scowcroft, W. R., and Gibson, A. H. (1975). Nitrogen fixation by *Rhizobium* associated with tobacco and cowpea cell cultures. *Nature 253,* 351-352.

Stewart, W. D. P. (1966). *Nitrogen Fixation in Plants.* Athlone Press, London.

Vela, G. R. Cagle, G. D., and Holmgren, P. R. (1970). Ultrastructure of *Azotobacter vinelandii. J. Bacteriol. 104,* 933-939.

Whitehead, T., McIntosh, T. P., and Findlay, W.M.(1953). *The Potato in Health and Disease.* Oliver and Boyd, Edinburgh.

Isolation of Biochemical Mutants of Cultured Plant Cells

Jack M. Widholm

Interest in obtaining plant cell biochemical mutants has been developing over the last few years. Such mutants might be indispensable for the successful completion of certain genetic modification experiments. Several plant cell culture mutants have been selected and some of these have been discussed previously (Widholm, 1974a). A short review would seem to be appropriate here, however.

Three general types of plant cell culture mutants have been found: auxotrophic, autotrophic and resistance. Auxotrophic mutants lack the ability to synthesize the necessary compounds needed for growth, presumably due to the lack of an enzyme. Autotrophic mutants no longer require a compound which is normally required. Resistance mutants are resistant to growth inhibition by compounds which are normally inhibitory.

Leaky auxotrophic mutants were selected by Carlson (1970) using 5-bromodeoxyuridine-induced suicide of mutagen-treated haploid *Nicotiana tabacum* cells. The cells that could grow on the minimal medium died due to the incorporation of 5-bromodeoxyuridine into their DNA, thus enriching the nongrowing auxotrophs. Of 119 colonies that grew on complete medium after enrichment from 1.75×10^6 treated cells, 6 were found which had partial requirements for specific compounds. Lines were identified whose growth could be specifically stimulated by hypoxanthine, biotin, P-amino-benzoic acid, arginine, lysine and proline. The requirements were not absolute since the lines grew without additions at rates of 16 to 50 percent of those obtained with supplementation. This leakiness was apparently caused by the fact that *N. tabacum* is an amphidiploid and not a true diploid therefore, haploid lines could still carry more than one copy of each gene. To select nonleaky auxotrophs from such lines may be difficult.

Hormone autotrophic lines have been described by several workers, but in most cases it is difficult to justify the use of "mutant" in describing the lines. Cells in culture should carry the genetic information coding for the enzymes required to synthesize auxin and cytokinin since plants are capable of such synthesis. In culture, however, most cell lines require for growth an exogenous source of both hormones.

Sacristan and Wendt-Gallitelli (1971) found a *Crepis capillaris* line which grew without exogenous auxin and cytokinin. These hormones were required by the normal strain. Plants were regenerated and cultures reinitiated. The new cultures reinitiated from plants grown from the hormone-autotrophic line did require hormones for growth. This indicated that the original autotrophy was not due to a stable mutation, but presumably was due to an alternate differentiation pattern.

One example of auxin-autotrophy which could be due to a mutation was described by Lescure and Peaud-Lenoel (1967). They selected sycamore cells in suspension culture which required no exogenous auxin. These lines were stable during culture and contained an altered indoleacetic acid oxidase which showed abnormal reaction kinetics (Lescure 1970). This could indicate that an alteration in auxin degradation rather than synthesis can cause autotrophy. The frequency of autotrophy was increased from zero to about 10^{-6} by mutagen treatment.

Several resistance mutants have been described. Heimer and Filner (1970) selected suspension cultured *N. tabacum* cells resistant to growth inhibition by threonine. Threonine inhibited normal cell growth by preventing nitrate uptake, but the resistant cells were able to take up nitrate in the presence of threonine. Resistance was stable for 40 cell generations of growth in the absence of threonine and was found in the cell population at a frequency of about 10^{-7} (P. Filner, personal communication).

Streptomycin-resistant petunia (Binding *et al.*, 1970) and tobacco (Maliga *et al.*, 1973a) mutants have been selected from haploid callus placed on inhibitory agar medium (0.5 to 1 mg/l streptomycin). Maliga and co-workers (1973a) found a resistance frequency of 10^{-6} for one tobacco variety and the trait was stable when passed through the plant stage. Maternal inheritance was indicated by plant crossing experiments, indicating that altered 70s ribosomes could be the site of resistance-as can occur with some microorganisms.

Maliga and co-workers (1973b) also described an *N. tabacum* line resistant to 5-bromodeoxyuridine. While no mechanism of resistance has been reported, the resistance was stable and was found with a frequency of about 10^{-6}.

Ohyama (1974) also has selected 5-bromodeoxyuridine-resistant lines, in this case using soybean protoplasts in liquid medium. The resistant cells were found with a frequency of 4×10^{-5} in nitrosoguanidine treated cells, but controls were apparently not included to determine the frequency in the non-mutagenized population. The resistance mechanism is unclear since both normal and resistant lines incorporate 5-bromodeoxyuridine into their DNA. This should be lethal if the cells are grown in light.

Carlson (1973) isolated methionine sulfoximine-resistant lines from mutagen-treated haploid *N. tabacum* cells which were plated on inhibitory medium. Three plants were regenerated from the resistant cells and two of these contained 5-times the normal levels of free methionine, indicating that resistance might be due to an oversynthesis of methionine.

Chaleff and Carlson (1974) have selected rice cells resistant to a lysine analog (S-2-aminoethyl-L-cysteine), by plating mtuagen-treated cells on medium containing inhibitory levels of this analog. Three lines were recovered which grew slowly in the presence of S-2-aminoethyl-L-cysteine. These lines contained about twice as much free lysine as normal and alterations in various other amino acids also were found. Enzyme studies were not carried out.

Our work with plant cell culture mutants has involved selecting diploid carrot and tobacco suspension cultures resistant to growth inhibition by amino acid analogs. To date, we have selected lines resistant to trytophan, phenylalanine, lysine, methionine and proline analogs. In each case, some of the lines oversynthesize the corresponding natural amino acid.

The selections were carried out by inoculating suspension cultured cells into liquid medium containing the analog at concentrations 2 to 5-times that which completely inhibited growth during short time incubations. This inhibitory concentration was determined by growing cells in various analog concentrations for a period of 10 days, after which the cell fresh weights were determined. The effect of P-flourophenylalanine on carrot growth is shown in Fig. 1. Studies were also carried out to determine if the inhibitory effect of the analog could be prevented by addition of the corresponding natural amino acid to the medium. The inhibitory effect of the analogs we have used for selection were usually prevented by the corresponding natural amino acid. This was important since the selection of amino acid overproducing lines should only be possible in cases where the analog growth inhibition can be prevented by the normal amino acid.

Cells in the inhibitory medium were incubated for up to two months-sufficient time for resistant cells to grow significantly. If growth occured in any flask, these cells were reinoculated into fresh inhibitory medium where growth usually continued. Mutagens were not used in the following selection experiments.

The first resistant lines selected and studied were tobacco and carrot cells resistant to 5-methyltryptophan (5MT) (Widholm 1972a, b, 1974a, b). Growth inhibition of normal tobacco and carrot cells in suspension culture was found to be complete at 14 and 44 μM 5MT, respectively. This inhibition was prevented by the addition of tryptophan, anthranilate, or indole, which are intermediates in the tryptophan biosynthetic pathway. Resistant tobacco and carrot lines were selected by incubating 3 x 10^6 and 6 x 10^6 cells in 100 ml of 44 and 220 uM 5MT medium, respectively, for up to two months. All flasks showed growth

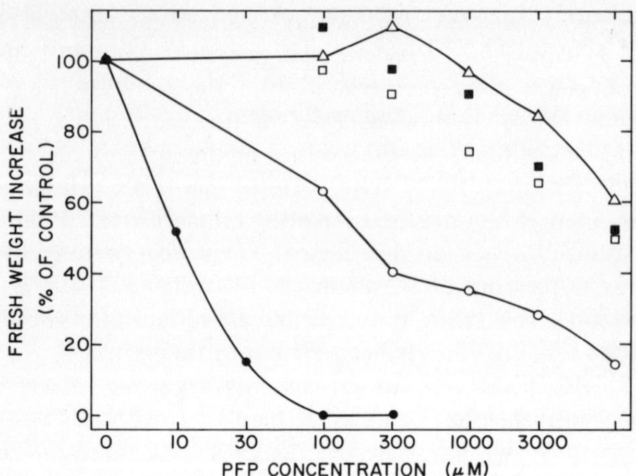

Fig. 1. The effect of p-flourophenylalanine (PFP) on carrot cells. Growth was measured as the increase in fresh weight after 8 days from an initial inoculum of 0.5 g fresh weight cells in 100 ml liquid medium. Senstive cells (●——●), resistant cells (Δ——Δ), resistant cells after having been grown for 50 cell mass doublings on a medium lacking PFP (○——○), average growth of 14 single cell clones isolated from the resistant line and grown away from PFP for 50 generations before the growth experiment (■——■), and the cloned lines grown on 328 uM PFP for 5 generations before the growth experiment (□——□). The fresh weights of the control flasks (no PFP) after the 8 day growth period were 10.1, 8.7, 8.2, 9.5, and 9.5 g, respectively. Reprinted with permission from Plant Physiology (Palmer and Widholm, 1975).

during this period, but when 10ml aliquots of the same cell and inhibitor concentrations were incubated, the frequencies of appearance of resistant cells could be estimated to be 1.5×10^{-6} for tobacco and 3×10^{-7} for carrot. Mutagen treatment did increase the mutation frequency in carrot cells.

The selected lines were resistant to over 100-times as much 5MT as were the normal cells and also were resistant to several other tryptophan analogs. The trait was stable for at least 60 to 100 generations. Recent unpublished experiments show that only 2 of 6 single-cell clones isolated from the 5MT-resistant carrot line were resistant to growth inhibition by 5MT. These clones retained the resistance, however, for over 100 generations, indicating that the original line continued to carry normal cells during growth on the inhibitor.

Both the resistant carrot and tobacco lines had a tyrotophan biosynthetic control enzyme, anthranilate synthetase, which was more resistant to feedback inhibition by tryptophan or 5-methyltryptophan. As a result of this altered control, both lines accumulated about 30-times more free tryptophan than normal. Thus, an altered feedback control enzyme allowed an oversynthesis which caused resistance to an analog.

During the last four years we have selected about 2000 5MT-resistant carrot lines and about 100 5MT-resistant tobacco lines. Of approximately 100 carrot and 30 tobacco lines which have been studied, almost all were stably resistant and had an altered anthranilate synthetase.

One 5MT-resistant carrot line was isolated and described which did not have an altered tryptophan control enzyme (Widholm 1974a). The resistance of this line appeared to be due to decreased 5MT uptake. The trait was stable during passage from culture to regenerated plants and back into culture.

Tobacco and carrot lines resistant to the phenylalanine analog, P-flouro-phenylalanine (PFP), have also been selected (Palmer and Widholm, 1975). The growth of normal carrot cells was completely inhibited by 100 μM PFP (Fig. 1). This inhibition could be partially reversed by 300 uM phenylalanine. Long term incubation of carrot cells in 328 uM PFP yielded one line which was resistant. This line was still able to grow in PFP concentrations 100-times those which inhibited normal cell growth (Fig. 1). When the resistant line was grown for 50 generations away from PFP, some loss of resistance was noted. All 14 single cell clones isolated from the resistant line maintained their resistance, however, when grown away from PFP for 50 generations. Growth for 5 generations with PFP did not change the resistance of these clones (Fig. 1). Three of these clones have been grown for 75 additional generations away from PFP and have maintained their resistance. A tobacco line resistant to PFP was selected in a similar way and it required about 10-times higher PFP concentrations for growth inhibition than did normal cells.

The PFP-resistant tobacco and carrot lines incorporated less ^{14}C-PFP into protein than did normal cells. This decrease was partially due to decreased uptake, but was also affected by an enlarge metabolic pool in both lines. The resistant carrot lines contained 6.3-times the normal level of free phenylalanine. The tobacco line had an altered control enzyme, chorismate mutase, which was more active and was more resistant to feedback inhibition (Fig. 2). These enzyme alterations should have caused the oversynthesis and accumulation of phenylalanine in the tobacco cells. Instead, the phenylalanine was apparently converted to phenolic compounds which did accumulate to levels of 6-fold higher than normal.

Recently, suspension cultured tobacco lines have been selected which were resistant to the lysine analogs, delta-hydroxylysine and S-2aminoethyl-Lcysteine. The 2 resistant lines required about 50-times as much of the analogs to inhibit growth and each line also accumulated more than 10-times as much free lysine as did the normal cells.

A carrot line has been selected which was resistant to ethionine, a methionine analog. Over 500-times as much ethionine was required to inhibit the growth of the resistant line. This resistance arose at a very low frequency,

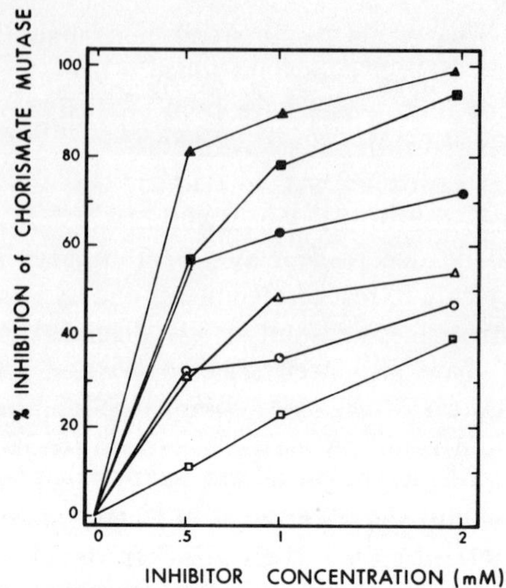

Fig. 2. Inhibition of chorismate mutase in crude extracts from tobacco cells by DL-p-flourophenylalanine (PFP) (■,□), L-phenylalanine (●,○) and L-tyrosine (▲,△). Closed symbols represent activity from sensitive cells and open symbols from resistant cells. Results are the means from two experiments. The sensitive cell extracts produced 72 and 87.6 nmoles prephenate in 30 min with 0.45 and 0.35 mg protein while the resistant cell extracts produced 60 and 70.8 nmoles with 0.12 and 0.16 mg protein with no inhibitors present. (Reprinted with permission from Plant Physiology (Palmer and Widholm, 1975).

was stable for at least 50 generations and the cells accumulated more than 10-times the normal levels of free methionine.

Another carrot line has been selected for resistance to hydroxyproline. This line accumulated about 30-times as much free proline which remained high for at least 50 generations after removal of hydroxyproline from the medium. The resistance was likewise retained.

These lysine, methionine and proline analog resistant lines have not been studied thoroughly enough to determine why the corresponding natural free amino acids accumulate. It would be logical to expect that a control mechanism has been altered. Study of these enzymes could give a basic understanding of control mechanisms in the affected biosynthetic pathways. These mutants also show that nutritionally important amino acids like tryptophan, lysine and methionine can be increased in plant cells.

The amino acid overproducing lines were selected from cell lines which had been in culture for several years and would no longer regenerate plants. Attempts are now being made to select similar mutants cell lines capable of plant regeneration.

In several of the mutant selection studies I have discussed, the authors reported that mutagens were used, but in no case was evidence given to show that this treatment had any effect on the mutation frequency (Carlson, 1970, 1973; Chaleff and Carlson, 1974; Heimer and Filner, 1970; Ohyama, 1974). In fact, Heimer and Filner (1970) said that N-methyl-N'-nitro-N-nitrosoguanidine (NG) treatment was not necessary for the recovery of mutants. Mutagens were not used by Maliga and co-workers (1973a, 1973b) in the selection of tobacco cells resistant to streptomycin or 5-BUdR and mutants were recovered in some cases with a frequency of 10^{-6}. Our selections were likewise carried out without mutagen treatment.

There is however, one reported case of mutagen treatment being necessary for the selection of cultured cell mutants. Lescure and Peaud-Lenoel (1967) selected sycamore cells which were auxin autotrophic; that is, the cells did not require the addition of an auxin to the medium for growth. Exponentially growing suspension cultured sycamore cells were treated with 400 g/ml NG for five min with no agitation. The cells were filtered and incubated in fresh medium which contained 1 mg/l 2, 4-D for eight days to allow some recovery of vigor. The cells were then screened to select clumps of less than 50 cells (50-200u) and incubated in liquid medium lacking auxin. After six weeks, clumps of green cells were noted only in the flasks containing NG-treated cells and not untreated cells. These growing cells continued to grow in medium lacking auxin.

From one to ten clumps grew in each flask of the selection medium (50 ml) which contained 1.5 x 10^7 cells. Only 10% of the originally inoculated cells were alive as ascertained by ability to grow in complete medium. The mutation frequency was thus increased from 0 to about 10^{-6}.

The mutants were stable and contained an altered IAA oxidase (Lescure 1970) which showed different reaction kinetics when compared with the enzyme from normal cells. This altered enzyme apparently allowed sufficient accumulation of endogenously synthesized auxin for the cells to grow without exogenous auxin.

Our work with mutagens was done with suspension cultured carrot cells which were treated with either ethylmethane sulfonate or ultraviolet light. Cells were incubated with 0.25% ethylmethane sulfonate for two to three hours after which the chemical was rinsed away with fresh medium. Other cells were irradiated with ultraviolet light in open petri dishes and then inoculated into fresh medium. After 24 hours, the treated cells were examined under a microscope-using phenosafranine which stains dead cells (Widholm 1972c)-to determine viability. The treatments killed 50 to 70% of the cells while 95% of the untreated cells were alive. To dilute out dead cells, the viable cells were allowed to grow until the cell mass had increased 4- to 8-fold. At this time, 1.8 x 10^5 cells were inoculated into 10 ml of 5MT inhibitory liquid medium, incubated for two months, and growth noted.

The data in Table 1 were obtained from several unpublished experiments. There was growth in 2 of 63 flasks of untreated carrot cells, giving a frequency of about 1.8×10^{-7}. Ethylmethane sulfonate treated cells grew in 42 of 70 flasks for 3.3×10^{-6} frequency or a 20-fold increase, assuming that there was one resistant cell in each flask in which growth occurred. Ultraviolet light treatment gave 22 of 70 for a frequency of 1.7×10^{-6} or a 10-fold increase over the untreated cells. Thus, the mutagens ethylmethane sulfonate and ultraviolet light substantially increased the 5-methyltryptophan resistance frequency.

Table 1. Effect of ethylmethane sulfonate and ultraviolet light on the frequency of 5-methyltryptophan-resistance in carrot cells.

Treatment*	Growth detected Number of flasks Exhibiting Growth	Resistance frequency
Control (no treatment)	2/63	1.8×10^{-7}
Ethylmethane sulfonate	42/70	3.3×10^{-6}
Ultraviolet light	22/70	1.7×10^{-6}

**The cells were treated and incubated as described in the text and the number of flasks showing growth were recorded. The calculated resistance frequency assumed that there was one resistant cell present in each flask in which growth was detected. Each flask contained 1.8×10^5 cells (10 mg fresh weight).*

This discussion points out the fact that only a few mutants from cell culture have been selected and characterized. Only one report concerning auxotrophic mutant selection attempts is in the literature. I believe that much more work will be done in the future and these mutants will have many varied and important uses. Uses would include growing plants with increased free amino acid levels and model systems for transformation and protoplast fusion experiments.

REFERENCES

Binding, H., Binding, K., and Straub, J. (1970). Selektion in gewebekulturen mit haploid Zellen. *Naturwissenschaften 57*, 138-139.

Carlson, P. S. (1970). Induction and isolation of auxotrophic mutants in somatic cell cultures of *Nicotiana tabacum. Science 168*, 487-489.

Carlson, P. S. (1973). Methionine sulfoximine-resistant mutants of tobacco. *Science 180*, 1366-1368.

Chaleff, R. S., and Carlson, P. S. (1974). *In vitro* selection for mutants of higher plants. In *Modification of Information Content of Plants* (R. Markam, ed.) John Innes *Symposium*

Heimer, Y. M., and Filner, P. (1970). Regulation of the nitrate assimilation pathway of cultured tobacco cells. II. Properties of a variant line. *Biochim. Biophys. Acta 215,* 152-165.

Lescure, A. M. (1970). Preparation de clones mutants de cellules vegetales. Recherche de 1' impact moleculaire de la mutation chez une lignee mutante d'acer pseudoplatanus L. independante de l'auxin. *Bull. Soc. Chimie Biol. 52,* 953-978.

Lescure, A. M., and Peaud-Lenoel, C. (1967). Production par traitement mutagene de ligness cellulaires d' acer pseudoplatanus L. anergiees a l'auxine. *Comp. Rendus Acad. Science Paris 265,* 1803-1805.

Maliga, P., Sz.-Breznovits, A., and Marton, L. (1973a). Streptomycin-resistant plants from callus culture of haploid tobacco. *Nature New Biology 244,* 29-30.

Maliga, P., Marton, L., and Sz-Breznovits, A. (1973b). 5-bromodeoxyuridine-resistant cell lines from haploid tobacco. *Plant Science Letters 1,* 119-121.

Ohyama, K. (1974). Properties of 5-bromodeoxyuridine-resistant lines of higher plant cells in liquid culture. *Exp. Cell Res. 89,* 31-38.

Palmer, J. E., and Widholm, J. M. (1975). Characterization of carrot and tobacco cell cultures resistant to P-flourophenylalanine. *Plant Physiol. 56,* 233-238.

Sacristan, M. D., and Wendt-Gallitelli, M. F. (1971). Transformation to auxin-autotrophy and its reversibility in a mutant line of *Crepis capillaris* callus culture. *Mol. Gen. Genet. 110,* 335-360.

Widholm, J. M. (1972a). Cultured *Nicotiana tabacum* cells with an altered anthranilate synthetase which is less sensitive to feedback inhibition. *Biochim. Biophys. Acta 261,* 52-58.

Widholm, J. M. (1972b). Anthranilate synthetase from 5-methyltryptophan-susceptible and resistant cultured *Daucus carota* cells. *Biochim. Biophys. Acta 279,* 48-57.

Widholm, J. M. (1972c). The use of flourescein diacetate and phenosafranine for determining viability of cultured plant cells. *Stain Technology 47,*189-194.

Widholm, J. M. (1974a). Selection and characteristics of biochemical mutants of cultured plant cells. In *Tissue Culture and Plant Science 1974* (H. E. Street, ed), pp. 287-299. Academic Press, New York.

Widholm, J. M. (1974b). 5-methyltryptophan-resistance trait carried from cell to plant and back *Plant Science Letters 3,* 323-330.

The Applicability of Plant Cell and Tissue Culture Techniques to Plant Improvement

Oliver E. Nelson, Jr.

With the advent of techniques making it feasible with some species of plants to handle the material alternatively as an assemblage of haploid cells growing in a defined culture medium or as a typical sporophyte has come the realization that this ability to manipulate the cultured cells in a manner analogous to a population of microbial cells might offer important advantages for plant improvement (Carlson, 1973a). Specifically, these advantages lie in the large populations that can be handled *in vitro*, the ease of mutagen application, the desirability of using haploid cells in mutation induction and the possibility of utilizing selective screens for detecting desired variants that subsequently can be recovered as diploid plants.

Such an approach constitutes an intellectually stimulating possibility for plant improvement, linking, as it does in theory, the subject to the techniques that have been so successful in molecular biology over the past two decades. Indeed, there are some who feel that such techniques have relieved plant science investigators of all constraints in the manipulation of plant systems. The purpose of this paper is to assess the possible utility of molecular genetics in plant improvement from the viewpoint of an interested and sympathetic observer. Such an assessment may assist those who are responsible for the allocation of research support to a new and potentially important area. Another discussion of plant cell culture techniques in relation to the improvement of crop plants has recently appeared (Carlson and Polacco, 1975).

At the present time, we have not advanced beyond some intriguing demonstrations (Nitsch, 1974; Carlson *et al.*, 1972; Carlson, 1973a; Carlson, 1973b; Doy *et al.*, 1973) and predictions that we have techniques at hand (or soon may have) that could be immensely useful in plant improvement. Sober appraisals of the possible utility of molecular genetics techniques by different scientists may well be quite divergent depending on the assumptions made as to which of the present barriers to use of these techniques are technical and can be expected to be surmounted within reasonable time spans and which represent true intractability of the biological system to manipulation. This situation can be expected to pertain for a number of years until research results delineate the areas where concerted efforts have not been successful.

It is useful to separate the techniques under discussion into three categories: those concerned with the cycle from diploid sporophyte through haploid cells in culture and back to diploid sporophyte; those concerned with the introduction of exogenous DNA into cells or plants with the hope of perpetuation and expression of the information therein; and the production of plant protoplasts that can be used to effect the fusion of cells from different species and in the most favorable cases, to enable the formation of interspecific hybrids. Each category, if perfected, encompasses techniques that conceivably could be useful in plant improvement. The three sets of techniques used in conjunction could be a most powerful tool. Nevertheless, the attempted use of one set of techniques may be independent of use of the other two. The first set of techniques appears to have the best prospect of utilization in the near future, and discussion will focus on this category although not to the exclusion of the others.

In attempting to evaluate the potential value of the techniques of molecular genetics, I assume the totipotency of most plant cells. Given the methods of producing haploid cells, culturing these and applying selective screens following mutagenesis, evoking callus and ultimately plantlet formation, one sees the outline of possible procedures for plant improvement. Owing to a variety of imponderables involving our ability to carry out each of these steps for a given species, I will adopt a "best possible case" estimate in attempting to assess the usefulness of these techniques. I assume the ability to carry out each of the above steps for any species of interest. At the present time, this ability exists for relatively few species although recent advances have added capabilities for several species in one or more of these areas. It is now possible, for example, to regenerate maize plants from callus cultures (Green and Phillips, 1975). It is further assumed that the product of applying these procedures is to be an improved line or cultivar of a species to be used in conventional agricultural systems, thereby ruling out considerations of tissue cultures from higher plants being used in unconventional production techniques.

To date, the impact of molecular genetics techniques on plant improvement has been negligible. Even with *Nicotiana tabacum*, the most easily manipulated species through the steps from haploid cells in culture to diploid sporophytes, the production of cultivars via production of haploid plantlets arising from anther cultures (which is possibly the simplest use of the described techniques) may require another four to five years (Collins and Legg, 1974). This method requires shorter periods of time to produce new homozygous genotypes than conventional breeding methods. This factor may not, in itself, be sufficient that all tobacco breeding will be carried out on this basis.

It is instructive to remember that while inbred lines of maize are ordinarily produced by five or six generations of self-pollination, it has been known for twenty-five years that inbred maize can be produced in a single step by selecting the rare spontaneously occurring haploid seeds and effecting a doubling in chromosome number prior to self-fertilization (Chase, 1969). Although one might assume this technique to be a signal advance that would revolutionize maize breeding (and it is clearly ingenious), the technique has been relatively little used in the intervening years. The reason for the limited application of this method of maize breeding has been the fact that the rate-limiting step in breeding hybrid maize has been the evaluation of the combining ability of the inbreds and not their production. Employment of the haploid method of maize inbred production does not alleviate this limitation, and consequently, the technology is presently almost unused. Thus, a new technique that appears, *a priori,* to be a substantial advance may not succeed for quite prosaic reasons.

Given an organism amenable to the suggested experimental manipulations, what improvements might one hope to accomplish? In the first place, it should be possible to select for resistance to bacterial, fungal, or insect toxins, for inhospitality to virus multiplication, or for resistance to the action of a herbicidal compound. The production of tobacco plants resistant to the toxin of *Pseudomonas tabaci* by such a selective system has been reported (Carlson, 1973b). It should also be possible to select plants with greater efficiency in the uptake and/or utilization of inorganic nutrients provided these attributes have their basis in cellular characteristics and not in supracellular organization.The same possibilities exist for selecting variants that are less injured by inorganic ions that are deleterious to plant growth. It is known for some species that ability to grow normally on low levels of a particular nutrient is simply inherited. The ability of celery to grow normally on soils with low quantities of magnesium or low quanitites of boron is conditioned by two different dominant genes (Pope and Munger, 1953a, b). By contrast, snap bean (*Phaseolus vulgaris)* cultivars that are efficient users of potassium are homozygous recessive for a single gene (Shea *et al.*, 1967).

Selection for the alteration of any step in a biosynthetic sequence should be possible provided that those reactions are being carried on in the single cell or in callus tissue and that suitable selective screens can be devised. Some of these mutants may well be affecting the regulation of production of essential metabolites. A system has recently been described that should allow the selection of variants in which the synthesis of either lysine, threonine, or methionine is enhanced by the presumed release of feedback inhibition on the first enzyme of the pathway (Green and Phillips, 1974). With this system, selections might be made either in tissue cultures or in seedlings.

The possiblity of selection for differences in cellular organelles also can be foreseen although the techniques by which desirable variants could be selected remain to be established.

In attempting to assess the utility of cell culture techniques in plant improvement, it is important to consider not only what might be considered the potentialities of the method but also those areas of plant improvement where advances by cell culture techniques do not appear feasible. Selection for those traits that are expressed only on an organismal basis or in specialized tissues (that is as a result of differentiative events) could be achieved only in instances where an organismal trait followed as a natural consequence of a change that could be selected at the cellular level.

Clearly, the most important characteristics of economically valuable plants are the quality and quantity of the fiber or food for which the plant is grown. In many cases, these are specialized tissues (the endosperm of cereal grains for example) or are produced in specialized tissues or organs and hence, not amenable to selection at the level of cultured plant cells. It is difficult to imagine selection for or against a particular species of storage protein in cell culture since the indications are that such storage proteins are produced only in the endosperms of cereal grains or the cotyledons of legume seeds. Further, the genetic basis of yield is complex, being supported by desirable alleles at a large number of loci. Some of these components of yield produce qualitative changes (tall vs. dwarf plants or disease-resistant vs. disease-susceptible plants) but many others produce quantitative changes that can be recognized and selected only under conditions in which the other components of the system express themselves. It is precisely in this area that much plant breeding effort is concentrated. Many of the polygenic yield factors seem inaccessible to identification or selection in cell culture systems.

Although I've accepted as a basis for discussion the successful completion of the cycle from diploid plant through haploid cells to diploid plants, some difficulties present themselves as impediments to the most successful utilization of plant cell culture systems. The most important is aneuploidy which is known to develop in many plant cells being cultured (Chaleff and Carlson, 1974). It has long been known that mammalian cell cultures (with the exception of fibroblasts which will remain as diploid cells but divide only a limited number of times) may be quite variable in karyotype. But investigators using mammalian cells in culture as experimental objects have no hope of regenerating complete organisms from their cells and can tolerate karyotypic abnormalities for many purposes. Investigators of plant cell cultures often wish to regenerate plants, where aneuploidy would frequently be disruptive.

One can foresee other complications to selection in cell cultures for desired variants. Many variations may have their basis in epigenetic changes rather than genetic changes. The epigenetic changes may either result in the expression of genetic information that had not been expressed in the cell line or in the suppression of genetic information that had been transcribed and translated in the cell lines up to that point. Neither type of epigenetic change alters the genetic information present but only what information is being utilized in a cell and its progeny. Thus, variants stemming from epigenetic changes may be erased if plants go through meiosis. During meiosis all epigenetic signals should be erased and the regulatory situation returned to a ground state. Epigenetic changes can constitute false positive signals if one seeks mutational changes conditioning a particular phenotype.

A major item of discussion over the past few years among investigators of animal cell cultures has been the extent to which variants (particularly analog-resistant) selected in cell cultures are epigenetic rather than genetic in origin. On the basis of similar rates of change to azaguanine resistance in diploid, tetraploid and octoploid male Chinese hamster cells, it has been suggested that epigenetic changes constituted the basis of the variation (Harris, 1973). If mutation were the basis of the variation, the expectation is that the mutation rate to phenotypic resistance should be the mutation rate for a single copy of the locus raised to a power equivalent to the ploidy level.

The basis of resistance to azaguanine is the loss of hypoxanthine guanine phosphoribosyl transferase activity, which would be expected to be recessive. The structural gene for the enzyme is X-linked, and this reduces the ploidy level of the gene in male cells but does not alter the argument. It has been similarly suggested that BUDR-resistance in cultured frog cells has a non-genetic basis since the rate of spontaneous changes arising in haploid and diploid cell lines was not in accord with the expectations, assuming that the basis of resistance was either a dominant or a recessive mutation (Mezger-Freed, 1972). The evidence that some analog-resistant variants have a genetic basis has been summarized (DeMars, 1974). Recently it has been demonstrated that azaguanine resistance in Chinese hamster cell lines can arise from mutation plus chromosome elimination (Chasin and Urlaub, 1975).

The evidence is scanty and inconclusive as to what proportion of analog-resistant variants in animal cell cultures have a genetic basis, and each variant constitutes a separate case. It seems probable, however, that both epigenetic and genetic changes contribute as a cause of such variants and that selection for desired variants in plant cell cultures will isolate a number of selections in which the basis of the variation is epigenetic.

A possible example of such epigenetic variations being selected in plant tissue cultures may be found in recent research at the University of Wisconsin (Mok, 1975). Callus cultures were derived from the roots of red, dark orange, light orange, yellow and white carrot varieties. All cultures produced the carotinoid pigments typical of the variety from which they were derived, although in decidedly lower quantities. The quantities produced in tissue culture were proportional to the quantities in the roots of the varieties from which they were derived. The tissue cultures gave rise to various clones that obviously differed in carotinoid content. For example, red-orange, pink, and yellow cultures were isolated from the callus cultures of the red-rooted variety by subculturing spontaneously appearing sectors. These isolates all had lower carotinoid contents than the original. All cells observed in the tissue cultures (original as well as variant) were aneuploid with all having more than the diploid number of chromosomes (2n = 18). When plants were regenerated from the variant cultures, the plants in each case were typical of the variety from which the original culture was derived, and all plants were diploid. The carotinoid contents of roots from the regenerated plants were quite similar to that of the original variety. The original red variety had a total carotinoid content (ug/g) of 60 in the xylem tissue and 151 in the phloem. The plant regenerated from a pink isolate had a xylem tissue content of 58 and a phloem tissue content of 146. The plant regenerated from a yellow isolate had a xylem content of 61 and phloem content of 150.

In the above instance, the investigator selected for variation in a trait expressed both in tissue culture and in the roots of plants. Spontaneous variants were easily detected and propagated, yet the plants derived from them returned to the original type. It is not clear whether the basis of the spontaneous deviations was epigenetic change in the culture or the readily demonstrable aneuploidy. One may conclude either that diploid cells were present in all the isolates and it was from these that embryoid formation took place or that aneuploid cells were regulated back to a diploid state during embryoid formation. In any case, the experience indicates that some disappointments may be anticipated even when selection for a trait in cultured cells can clearly be effected. There are instances, however, in which complete positive correlation is observed between a trait in cultured cells and in the diploid sporophyte as shown for resistance to *Phytophthora parasitica* var *nicotianae* in tobacco (Helgeson *et al.*, 1976).

In addition to the cycle from diploid plants to haploid cells in culture and back to diploid plants, which does have obvious implications for plant improvement, two other possibilities are frequently mentioned as relevant to plant improvement, These are the transfer of genetic information from one species to another, either by a process akin to transformation in bacterial systems or by utilizing a vector (a virus or a plasmid that could replicate itself in plant cells) to carry the information. If methods existed for introducing desired genetic information into plant cells (or into a plant), insuring its integration or its perpetuation as an extra-chromosomal entity and the expression of the genetic information contained, clearly much of a discussion devoted to the influence of molecular genetics on plant improvement would deal with this issue. The methodology for effecting such transfers, however, does not presently exist.

It has been possible to demonstrate in an extensively studied system that DNA carrying a specific marker gene can be taken up by eucaryotic cells with subsequent expression of the information from the marker gene in a random fashion. There is, however, no integration of the introduced genetic information to supplant that existing in the host (Fox 1977). While the techniques of introducing foreign DNA into bacterial plasmids have reached a high degree of sophistication and the same techniques can be utilized for plants, there is a paucity of suitable vectors that could be utilized for plant cells.

Another technology much discussed in relation to plant improvement is protoplast fusion. The assumption is that such fusion between cells of different species, different genera, or different families may be a valuable tool in effecting hybridization where conventional crossing methods fail. It is possible to hybridize by protoplast fusion two species of *Nicotiana* that can be crossed by conventional means (Carlson *et al.,* 1972), and cells from plants of different families can be induced to fuse although the fusion product rarely divides more than once (Kao *et al.,* 1974).

I am not sanguine about the results of fusion between cells of plants that are widely separated taxonomically. It is difficult to conceive that two diverse developmental programs in a single nucleus (assuming that the two nuclei do fuse) can lead to any degree of successful differentiation. There are numerous instances known in which no barrier to fertilization exists but the zygote cannot develop successfully owing to developmental disharmonies, the basis of which are unknown. For this reason, I suspect that there will be relatively few instances in which it will be profitable to employ protoplast fusion to effect hybridization by evading incompatibility barriers. It is predictable that the fusion products between cells of widely separated species would be as developmentally impotent, as has been the situation to date. Possibly the most useful outcome of fusion experiments would be the progressive elimination of the chromosomes of one species as the cells divide. This frequently happens following the enforced fusion of cells from two different species of animals. If cell culture took place under selective conditions, it might be possible to retain a chromosome or a portion of a chromosome containing desired genetic information from the species whose chromosomes were being eliminated as has occurred in fusion products between the cells of two different species of animals. By irradiation, the desired genetic information could be transferred into the dominant genome, thus utilizing protoplast fusion as a means of effecting gene transfer. The genomes of the two species, however, must initially be sufficiently harmonious to allow cell wall formation and cell division, and this requirement may well limit the taxonomic separation of species on which it can be employed. The elimination of the chromosomes of one species under the apparent influence of the genome of another species has been observed in several interspecific crosses following conventional crossing techniques (Davies, 1974). Since this paper was given, it has been reported that culture on selective medium of cells that are the result of fusion between *Petunia* and *Parthenocissus* cells results in the elimination of the *Petunia* chromosomes although the hybrid cells retain isoperoxidases specific to the *Petunia* parent for periods as long as one year after which these isoperoxidases are progressively lost (Power *et al.*, 1975).

The mingling of cytoplasms from different varieties within a species or two closely related species that could occur via protoplast fusion conceivably would allow an opportunity for recombination of genes carried on organellar DNA. This would not occur in ordinary crosses of some species since chlorplasts are not transmitted by the male parent in most species of plants. Possibly such mingling of the cytoplasms or the organelles contained therein from two species could be beneficial in some instances. There is no substantial evidence that such interaction occurs although it has been suggested that the basis of hybrid vigor is the interaction of mitochondria from the two parents. (Sarkissian, 1972). Mitochondria are contributed to the zygote by the male parent in at least some plant species (Hoeffert, 1969). Observations of mitochondrial complementation between different varieties or inbreds have not been substantiated in another laboratory (Ellis *et al.*, 1973).

Then it is reasonable to ask: Where do these techniques derived from research in plant physiology and molecular genetics fit into programs in plant improvement? This is particularly appropriate since popular articles tend to depict the application of these techniques as the modern way to carry out plant breeding programs or molecular genetics answer to plant improvement. Since only several potential sugar cane varieties arising from tissue culture techniques could be construed in a broad sense as applications of newer technology to plant improvement (Nickell and Heinz, 1973) and since attempted applications to other economically important plants have not yet been successful, it is not surprising that many plant breeders are antagonized by suggestions that conventional methods of plant breeding are outmoded. They suspect that support for their programs will erode if such ideas become prevalent. It is important, both in terms of allocating research support and in inducing plant breeders to consider cooperative research with those involved in development of the technologies discussed here, to emphasize my conclusion concerning the utilization of these technologies. Even if these technologies become as successful as their most ardent proponents predict, orthodox plant breeding techniques will not be supplanted. New cultivars are not going to arise from the test tube or flask in many instances, if in any. Even the potential cultivars of tobacco developed by anther culture (which is the simplest utilization of these techniques) would require an extensive process of evaluation for chemical composition and for yield potentialities in comparison with the present elite cultivars. This is a process best carried on by plant breeders.

I visualize the successful application of the considered technologies as providing plant breeders with desirable variants that conceivably could not have been obtained by other means or could not have been obtained as economically in terms of space or time and which will then provide the basis for plant breeding programs. The efforts of molecular geneticists in this area will be complementary to plant breeding programs, just as the germ plasm collections of plant explorers are complementary contributions of additional desirable variations to breeding programs.

In view of the uncertainties surrounding the development of methodologies, it is difficult to suggest what degree of emphasis should be placed in plant improvement on the manipulation of plant cells in culture or molecular genetics experiments that might make it possible to transfer genetic information from one species to another. My appraisal suggests that the utilization of these technologies (even if successful) has definite limitations in scope. They are unlikely to allow selection for traits that arise as a consequence of differentiation or selection in the polygenic complexes that condition yield. These areas clearly constitute a major part of plant breeding efforts. For qualitative traits that are expressed in cell cultures, however, considerable experimental leverage is put at the disposal of the investigator. For this reason, I feel that we cannot afford not to invest sufficient time and resources to ascertain if the implied promise that it will indeed be a powerful tool for plant improvement can be fulfilled. The clearest indication that this is so will be the demonstration that an economically important plant (other than tobacco) can be improved in some desirable attribute by these techniques to an extent not possible by ordinary techniques or that the improvement can be effected in appreciably shorter time. Until then, the attitude of many plant breeders will continue to be one of interested skepticism.

REFERENCES

Carlson, P. S. (1973a). The use of protoplasts for genetic research. *Proc. Nat. Acad. Sci. (U.S.A.) 70,* 598-602.

Carlson, P. S. (1973b). Methionine sulfoximine-resistant mutants of tobacco. *Science 180,* 1366-1368.

Carlson, P. S., Smith, H. H., and Dearing, R. D. (1972). Parasexual interspecific plant hybridization. *Proc. Nat. Acad. Sci. (U.S.A.) 69,* 2292-2294.

Carlson, P. S., and Polacco, J. S. (1975). Plant cell cultures: genetic aspects of crop improvement, *Science 188,* 622 - 633.

Chaleff, R. S., and Carlson, P. S. (1974). Somatic cell genetics of higher plants. *Ann Rev. Genetics 8,* 267-278.

Chase, S. (1969). Monoploids and monoploid derivatives of maize *(Zea mays L.). Bot. Rev. 35,* 117-167.

Chasin, L. A., and Urlaub, G. (1975). Chromosome-wide event accompanies the expression of recessive mutations in tetraploid cells. *Science 187,* 1091-1093.

Collins, G. B., and Legg, P. D. (1974). The use of haploids in breeding allopolyploid species. In *Haploids in Higher Plants* (K. Kasha, ed.), pp. 231-247. University of Guelph, Guelph, Ontario.

Davies, D. R. (1974). Chromosome elimination in inter-specific hybrids. *Heridity 32,* 267-270.

DeMars, R. L. (1974). Resistance of cultured human fibroblasts and other cells to purine and pyrimidine analogues in relation to mutation detection. *Mutation Res. 24,* 335-364.

Doy, C. H., Greshoff, P. M., and Rolfe, B. M. (1973). Biological and molecular evidence for the transgenosis of genes from bacteria to plant cells. *Proc. Nat. Acad. Sci. (U.S.A.) 70,* 723-726.

Ellis, J. R. S., Brunton, C. J., and Palmer, J. M. (1973). Can mitochondrial complementation be used as a tool in breeding hybrid cereals? *Nature 241,* 45-47.

Fox, A. S. (1977). Gene transfer in *Drosophila melanogaster.* This volume.

Green, C.E., and Phillips, R.L. (1974). Potential selection system for mutants with increased lysine, threonine, and methionine. *Crop Sci. 14,* 827-830.

Green, C. E., and Phillips, R. L. (1975). Plant regeneration from tissue cultures of maize. *Crop Sci. 15,* 417-420.

Harris, M. (1973). Anomalous patterns of mutation in cultured mammalian cells. *Genetics 73 Supplement,* 181-185.

Helgeson, J. P., Haberlach, G. T., and Upper, C. D. (1976). A dominant gene conferring resistance to *Phytophora parasitica* var *nicotianae* is expressed in tobacco tissue cultures. *Phytopath. 66,* 91-96.

Hoeffert, L. L. (1969). Fine structure of sperm cells in pollen grains of *Beta. Protoplasma 68,* 237-240.

Kao, K. N., Constabel, F., Michayluk, M. R.. and Gamborg, O. L. (1974). Plant protoplast fusion and growth of intergeneric hybrid cells. *Planta 120,* 215-227.

Mezger-Freed, L. (1972). Effect of ploidy and mutagens on bromodeoxyuridine resistance in haploid and diploid frog eggs. *Nature, New Biology 235,* 245-246.

Mok, M. (1975). Carotinoid synthesis in tissue cultures of *Dacus carota L.,* Ph. D. Thesis, Univeristy of Wisconsin, Madison.

Nickell, L. G., and Heinz, D. J. (1973). Potential of cell and tissue culture techniques as aids in economic plant improvement. In *Genes, Enzymes, and Populations* (A. Srb, ed.), pp. 109-128. Plenum Press, N. Y.

Nitsch, C. (1974). Pollen culture - a new technique for mass production of haploid and homozygous plants. In *Haploids in Higher Plants* (K. Kasha, ed.), pp. 123-135. University of Guelph, Guelph, Ontario.

Pope, D.T., and Munger, H.M. (1953a). Heredity and nutrition in relation to magnesium deficiency chlorosis in celery. *Proc. Amer. Soc. Hort. Sci. 61,* 472-480.

Pope, D.T., and Munger, H.M. (1953b). The inheritance of susceptibility to boron deficiency in celery. *Proc. Amer. Soc. Hort. Sci. 61,* 481-486.

Power, J. B., Frearson, E.M., Hayward, C., and Cocking, E.C. (1975). Some consequences of the fusion and selective culture of petunia and Parthenocissus protoplasts. *Plant Science Letters 5,* 197-207.

Sarkissian, I. V. (1972). Mitochondrial polymorphism and heterosis. *Z. Pflanzenzuchtung 67,* 53-64.

Shea, P. F., Gabelman, W. H., and Gerloff, G. C. (1967). The inheritance of efficiency in potassium utilization in snap beans (*Phaseolus vulgaris L.*) *Proc. Amer. Soc. Hort. Sci. 91,* 286-293.

Transducing Viruses and Viral Integration: Techniques for Genetic Modification

Kazunori Shimada, Robert A. Weisberg, and Max E. Gottesman

The purpose of this article is to review the current status of lambda-promoted transduction, with a view towards using transduction as a means of obtaining specific bacterial genes in high yields on a convenient virus vector.

THE INTEGRATION AND EXCISION OF BACTERIOPHAGE LAMBDA

The isolation of transducing lambda (λ) phage, that is phage which carry bacterial DNA covalently linked to the viral genome, usually requires integration of the phage DNA into the host chromosome. Integration results from reciprocal recombination between lambda DNA and *E. coli* DNA. The integrated phage genome is covalently linked to the host chromosome (Gottesman and Weisberg, 1971).

Lambda normally integrates at only one site or locus on the *E. coli* genome termed *att B.* This site is located between the galactose *(gal)* and biotin *(bio)* operons. The distinctive features of the base sequence of the *att B* site responsible for this specificity are not known. What is clear is that there is little homology between the base sequence of the DNA at the *att B* site and lambda DNA. Attachment involves the recognition of the specific DNA base sequences rather than simply the recognition of complementary base sequences. This is consistent with the observation that phage DNA integration requires a protein for integration (termed Int) encoded for by the phage genome. The Int protein does not promote general recombination, that is recombination between homologous DNA molecules that do not contain attachment sites, nor do the bacterial host general recombination functions promote normal lambda DNA integration.

The excision of prophage lambda from the *E. coli* chromosome is an efficient recombination process that requires two proteins encoded in the phage genome, Int and an excision protein (termed Xis) (Echols, 1970). Xis, like Int, is a site-specific and not a general recombination function.

Phage lambda integration and excision can now be observed *in vitro* and we expect rapid clarification of the biochemical mechanisms of these reactions.

THE FORMATION OF TRANSDUCING PHAGE

Lambda DNA site-specific recombination is almost always precise. A cycle of integration and excision alters neither the DNA of the phage nor of the bacterial host. Occasionally (about one in a million excision events) an error in excision occurs and a lambda DNA molecule is produced which has replaced a portion of its genome with bacterial DNA. These "transducing" phages carry bacterial genes as a permanent part of their genome and can transfer these genes from one bacterial host to another. The functions involved in the formation of transducing phage are unknown but the process is not believed to involve recombination between homologous DNA sequences.

The bacterial DNA acquired by the transducing phage particle derives from portions of the *E. coli* chromosome adjacent to the integrated prophage. The amount of bacterial DNA that can be packaged in the phage head is limited, in the usual case, to a maximum of about 16×10^6 amu or 25 kilobases. Thus, *E. coli* genes further than about ½ minute away from the prophage on the *E. coli* genetic map (Taylor and Trotter, 1972) normally cannot be transduced by lambda.

If the bacterial gene is close to the prophage integration site the loss of viral DNA need not result in a defective phage. The incorporation of the more distal *E. coli* genes into virions requires a greater compensatory loss of lambda DNA. Such transducing phage, λdgal, for example, have lost vital viral genes and are defective; they can only be grown to high titers or concentrations in the presence of normal lambda particles. Both plaque-forming and defective transducing phages serve as vectors of bacterial genes. The former, however, are more easily grown and purified and are therefore preferable.

EXTENDING THE RANGE OF LAMBDA TRANSDUCTION

As already stated, lambda (or its close relative, the phage φ80, with its attachment site near *trp*) can transduce only genes very near its attachment site. To transduce a distal bacterial gene, one must either (i) transpose the gene to a position near the prophage or (ii) establish the prophage near the distal gene.

Gene Transposition

The first technique, gene transposition, has proved quite successful in the preparation of a variety of novel transducing phage.

Transposition is achieved by the use of a bacterial F^1 episome, carrying the bacterial gene of interest. The episome is inserted, by selective pressure, into the bacterial chromosome near a prophage attachment site. Insertion of an episome whose autonomous replication is thermolabile is selected for by growth of the culture at non-permissive (usually high) temperature; the inserted F^1 is replicated as a part of the bacterial chromosome. The basis of the selection is that bacteria that have not inserted their F^1 episome die, since the F^1 carries some vital function.

Insertion can be directed to a particular site by in addition selecting for the loss of a bacterial function. In particular, episomal insertion into the *tonB* gene, located near the attachment site of φ80, disrupts and inactivates the gene. As a consequence, the bacterium becomes resistant to the bacteriophage T1, and can be selected for as a survivor of T1 infection.

A related technique (Press *et al.,* 1971) is based upon the phenomenon of episome exclusion - the inability of *E. coli* cells to stably maintain more than one F^1 episome simultaneously. Introducing two F^1 episomes into a cell, then selecting for functions carried by each, results in the isolation of cells in which the episomes have fused. If one episome carries a lambda attachment site and the other a distal bacterial gene, episome fusion approximates the bacterial gene to the attachment site.

Once a bacterial gene has been brought near a prophage attachment site isolation of a phage transducing this gene is fundamentally no different from isolating, for example, λ*gal* or λ*bio* transducing phage.

Establishment Of Prophage Near The Distal Gene

The second method of isolating novel transducing phage was developed by our group at the National Institute of Health (Shimada *et al.,* 1972, 1973, 1975). The possibility of translocating prophage lambda to different loci on the bacterial chromosome arose from an initial observation that deletion of *attB* reduced, but did not eliminate, lambda integration. Our studies on this Int-dependent residual integration suggested a potentially powerful method for the generation of a large variety of lambda transducing phage.

Integration of lambda promoted by the phage site-specific recombination function, Int, in the absence of the bacterial *attB* site, indicated that sequences resembling the normal attachment site occurred elsewhere in the *E. coli* genome. The number and location of these sequences is, of course, critical if the method is to be generally applicable. A simple selection for lambda lysogens by immunity to lambda superinfection revealed a number of different secondary attachment sites. The sites were not identical since there were repeated insertions at some sites and only occasional insertions at others. These lysogens have yielded several new and useful transducing phage; for example, λ*galR* and λ*dthy.*

An additional useful feature of secondary-site lysogens is their low efficiency of normal prophage excision. The frequency of excision errors is, however, not changed, with the result that the ratio of transducing to non-transducing phage in lysates is very high. This permits the isolation of plaque-forming transducing phage by their buoyant density (more or less than lambda) or by the loss of certain easily-identified lambda genes. Such transducing phage are normally difficult to observe among the much greater numbers of normal phage.

We are not limited, however, to transducing bacterial genes proximal to these more frequently utilized sites. In a manner entirely analogous to that described above for the directed insertion of F^1 episomes, the insertion of lambda can be directed by selecting for the loss of a bacterial function. The number of possible integration sites, when selective pressure is applied, is very large, occurring probably every few thousand base-pairs.

Proof that the loss of a bacterial function is the consequence of prophage insertion within the bacterial gene or operon is rather easier in this method than in the gene transposition method. Whereas gene destruction by F^1 is irreversible, lambda excision can restore gene function. Lambda lysogens can be "cured" of their prophage by brief derepression. Cured cells, easily detected, will show a return to wild-type phenotype in cases where the original loss of gene function was the result of lambda integration.

There are several techniques for selecting for loss of gene function. We mentioned above the use of T1 resistance to direct the insertion of F^1 episomes into *tonB* gene near the Ø80 attachment site. Lambda has also been used to inactivate genes encoding receptors for virulent phages. The *tsx* locus, encoding the receptor for T6, lies in an important region of the *E. coli* chromosome. Lambda integrates into *tsx* rather frequently and bacterial lysogens with lambda inserted at this locus have yielded transducing phage of considerable potential value (D. Friedman, personal communication). Similarly, the phage receptor locus *bfe* has served to direct lambda integration near the cistrons encoding the β and β^1 subunits of RNA polymerase, and phages transducing these cistrons have been isolated (Kirschbaum, 1973).

Other selections for mutants, provided that the method of selection is sufficiently sensitive to detect a mutant among about 10^6 cells (for example, penicillin enrichment) can probably be used to "direct" phage insertion. A newer technique for the isolation of auxotrophs uses the lambda prophage itself as a selecting agent (C. Dambly, personal communication). Lambda will not kill an auxotrophic host in minimal medium. To select for the integration of lambda into *trp,* for example, *E. coli* deleted for *attB* is infected with lambda and lysogens selected. The lysogens are then induced in the absence of tryptophan, effectively killing most *trp*$^+$ cells. The survivors, heavily enriched in *trp*-auxotrophs, are then tested by curing to verify the prophage location.

In some cases, a particular bacterial gene is located neither near *attB* nor near a frequently utilized secondary attachment site, nor near a marker which can be disrupted by lambda integration. Here, the gene itself may serve as a lambda insertion site. Thus, a lambda phage transducing part, or the whole of the *trp* operon, can be isolated from a lysogen containing lambda inserted in the *trp* operon. The methods for isolating phage carrying an intact copy of the disrupted gene are described in detail elsewhere (Shimada *et al.*, 1973, 1975) and we shall only briefly summarize them here.

An *E. coli trp*-lysogen isolated in this laboratory carried lambda integrated in *trpC*, which is in the middle of the *trp* operon. Transducing phage which transduced genes to one side of the prophage, λ*trpAB*, and to the other side λ*trpDE*, were obtained by the usual methods. These phage carried reciprocal fragments of *trpC*. To obtain phage carrying the entire operon, the two transducing phage were crossed. Site-specific recombination rejoined the disrupted *trpC* gene, just as prophage curing was able to restore the activity of a partitioned gene. The cross yields λ*trpABCDE*.

Alternatively, a phage which transduces half of the operon can be used to generate phage carrying the intact operon. λ*trpAB* will integrate at *trp* by homology, rather than at *attB* by site-specific recombination. From such lysogens, phage transducing the entire *trp* operon can be selected. These arise at the low frequency characteristic of the excision errors which generate transducing phage.

Recently, a technique for isolating a wide spectrum of transducing phage has been developed which does not require the directed integration of lambda (Schrenk, 1975). Instead, the technique depends upon the detection of very rare transducing phage among a large population of normal lambda. The method demands that the *E. coli* mutants to be transduced show a negligible rate of reversion. In the first step, *attB*-deleted *E. coli* are infected with lambda and random secondary-site lysogens are selected for by their immunity to lambda superinfection. A set of about 104 *E. coli* lysogens are induced, yielding a variety of different transducing phage representing the different secondary attachment sites in the set. By screening a number of sets, transducing phage for many bacterial cistrons have been obtained; the overall success rate is in excess of 50%.

Taken together, these methods of isolating transducing phage are probably capable of yielding phage carrying any bacterial marker desired. The transducing phage can be grown to high titer, resulting in a significant purification, of the bacterial gene relative to the *E. coli* chromosome. Further purification, if necessary can be obtained by selective DNA hybridization (Shapiro *et al.*, 1969). or by the use of restriction enzymes (Rambach and Tiollais, 1974; Thomas *et al.*, 1974). Mutagenesis of the transducing phage is also a convenient means of selectively altering a particular bacterial gene without affecting the remainder of the *E. coli* genome. Finally, lambda transducing phage have been reported to serve as vectors for the transmission of bacterial DNA to eucaryotic cells (Merrill, this volume). It appears, then, that lambda transducing phage will continue to play an important role in the future development of genetic engineering.

REFERENCES

Gottesman, M. E., and Weisberg, R. A. (1971). Prophage insertion and excision. In *The Bacteriophage Lambda* (A. D. Hershey, ed.), pp. 113-138. Cold Spring Harbor Laboratory, New York.

Gottesman, S., and Beckwith, J. R. (1969). Directed transposition of the arabinose operon: a technique for the isolation of specialized transducing bacteriophages for any *Escherichia coli* gene. *J. Mol. Biol. 44,* 117-127.

Guarneros, G., and Echols, M. (1970). New mutants of bacteriophage λ with a specific defect in excision from the host chromosome. *J. Mol. Biol. 47.* 565-574.

Kirschbaum, J. B. (1973). Regulation of subunit synthesis of *Escherichia coli* RNA polymerase. *Proc. Nat. Acad. Sci. (U.S.A.) 70,* 2651-2655.

Murray, N. E., and Murray, K. (1974). Manipulation of restriction targets in phage λ to form receptor chromosomes for DNA fragments. *Nature 251,* 476-481.

Press, R., Glandsdorff, N., Miner, P., DeVries, J., Kadner, R., and Maas, W. K. (1971). Isolation of transducing particles of ∅80 bacteriophage that carry different regions of the *Escherichia coli* genome. *Proc. Nat. Acad. Sci. (U.S.A.) 68,* 795-798.

Rambach, A., and Tiollais, P. (1974). Bacteriophage λ having *Eco*RI endonuclease sites only in the nonessential region of the genome. *Proc. Nat. Acad. Sci. (U.S.A.) 71,* 3927-3930.

Schrenk, W. J., and Weisberg, R. A. (1975). A simple method for making new transducing lines of coliphage λ. *Molec. Gen. Genet. 137,* 101-107.

Shapiro, J., MacHattie, L., Eron, L., Ihler, G., Ippen, K., and Beckwith, J. (1969). Isolation of pure *lac* operon DNA. *Nature 224,* 768-774.

Schimada, K., Weisberg, R. A., and Gottesman, M. E. (1972). Prophage lambda at unusual chromosomal locations. I. Location of the secondary attachment sites and the properties of the lysogens. *J. Mol. Biol. 63,* 483-503.

Shimada, K., Weisberg, R. A., and Gottesman, M. E. (1973). Prophage lambda at unusual chromosomal locations. II. Mutations induced by bacteriophage lambda in *Escherichia coli* K12. *J. Mol. Biol. 80,* 297-314.

Shimada, K., Weisberg, R. A., and Gottesman, M. E. (1975). Prophage lambda or unusual chromosomal locations. III. The components of the secondary attachment sites. *J. Mol. Biol. 93,* 415-429.

Taylor, A. L. and Trotter, C. D. (1972). Linkage map of Escherichia coli strain K12. *Bacteriol. Rev. 36,* 504 -524.

Thomas, M., Cameron, J., and Davis, R. W. (1974). Viable molecular hybrids of bacteriophage lambda and eukaryotic DNA. *Proc. Nat. Acad. Sci. (U.S.A.) 71,* 4579-4583.

Interactions of Bacterial Viruses and Bacterial Genes With Animal Systems

Carl R. Merril

The first interaction of bacterial viruses in mammalian systems was described by D'Herrele in 1917. These viruses were discovered in bloody stools of soldiers afflicted with dysentery. D'Herrele believed from his study of 34 dystentery patients that the bacterial viruses he had observed in these stools played a significant role in the recovery of the patients (D'Herelle 1917, 1918), therefore, he attempted to utilize these bacterial viruses as a treatment for bacterial diseases. This effort was cut short by the subsequent development of antibiotics and the bacterial viruses were relegated to serving as a cornerstone of modern molecular biology. Despite the large amount of research on bacterial viruses, their ecological role, particularly in relation to man, other mammals as well as their host bacteria is largely unknown.

Following completion of the bacterial virus course at Cold Spring Harbor during the summer of 1968, I decided to investigate whether bacterial viruses could interact directly with mammalian cells. One stimulus to investigate this area was the widespread practice of pipetting bacterial viruses by mouth. There seemed to be no doubt among the instructors and many of the students that direct interactions between mammalian cells and bacterial viruses were non-existent.

On searching the literature the only effect that could be found was a report of material isolated from either a lambda (λ) virus infected *E. coli* lysate or from disrupted lambda virus particles which could inhibit mammalian DNA viruses *(Herpes Simplex and Vaccinia)* in animal tissue culture (Centifanto, 1968). A confirmation of this report was later published in *Nature* (Meek and Takahasn, 1968). A similar ability of a virus T4 to interfere with the infectivity of mammalian viruses, in this case vascicular stomatitus virus (VSV), was reported (Kleinschmidt *et al.,* 1970) utilizing mouse L cells and live mice.

In February of 1969 we observed an inhibition of *Herpes Simplex* virus plaque formation on chick embryo fibroblasts by purified lambda. At the time, however, the effect was highly variable (Merril *et al.,* 1969). More recently we have found that a high multiplicity of infection of 10^5 - 10^6 with λ or ϕ80 virus (4 - 6 hours prior to infection with VSV) would inhibit VSV plaque formation (Merril and Roozen, 1974).

Searching for other direct effects of phage on mammalian cells, we looked for cytopathic phenomena. No reproducible effects were observed, although Mankiewicz (1965) had reported isolating phages that produce cytopathogenic effects in HeLa and African green monkey cells in culture. Although these experiments designed to detect phage replication were essentially negative, these studies eventually led to the discovery of widespread phage contamination in commercial fetal calf serum (Merril *et al.,* 1973).

Since the classic observational methods of virology employ a search for cytopathogenic effects, viral multiplication and/or cell death, these were looked for, but were not observed, with phage λ . It was decided to look for a possibly more subtle effect - the production of a single gene product.

By using galactosemic human skin fibroblasts which were deficient in the activity of a specific enzyme and the bacterial virus lambda-pgal (which carries the gene for the bacterial version of the defective enzyme), it was possible to test for a procaryotic gene function in human cells. The cells utilized were strain CCL-72 from the American Type Culture Collection. The first experiment utilized lambda virus and lambda virus DNA as control infections with λ pgal virus and λ pgal virus DNA in the experimental infections. The flasks were monitored by adding radioactive ^{14}C-labeled galactose at the time of the infection. Seventy-two hours after infection, the media was acidified by injecting 0.1 N HCl (0.1 ml). The flasks had been sealed with silicone rubber stoppers so that any radioactive gas produced would remain in the flasks. All of the gas was then flushed out of the flasks and through a trap containing phenethylamine absorbed to chromsorb W (Stuart and Williams, 1967). This material was then added to a toluene scintillation cocktail and counted. The first experiment was allowed to continue for an arbitrary 72 hours. We found twice as many counts in the flasks infected with λ pgal virus or its DNA than in the control flasks infected with lambda virus. The major difficulty with this experiment was the fairly high background or the large amount of carbon dioxide produced by the galactosemic cells.

The main pathway for galactose metabolism, both in procaryotes such as *E. coli* and mammalian cells, is the Leloir pathway as illustrated in Fig. 1. It was apparent from the experiment and subsequent experiments (Figs. 2, 3) that the alternate pathways of galactose metabolism were fairly efficient. We have continued our studies of carbon dioxide production from ^{14}C-galactose in galactosemic cells in an attempt to further elucidate the pathways and to eventually decrease the amount of carbon dioxide produced so that observations of the bacterial gene effect can be more clearly observed. We have found that the utilization of galactose in both normal and galactosemic cells was strongly inhibited in physiological concentrations of glucose (Petricciani *et al.,* 1972). The rate and

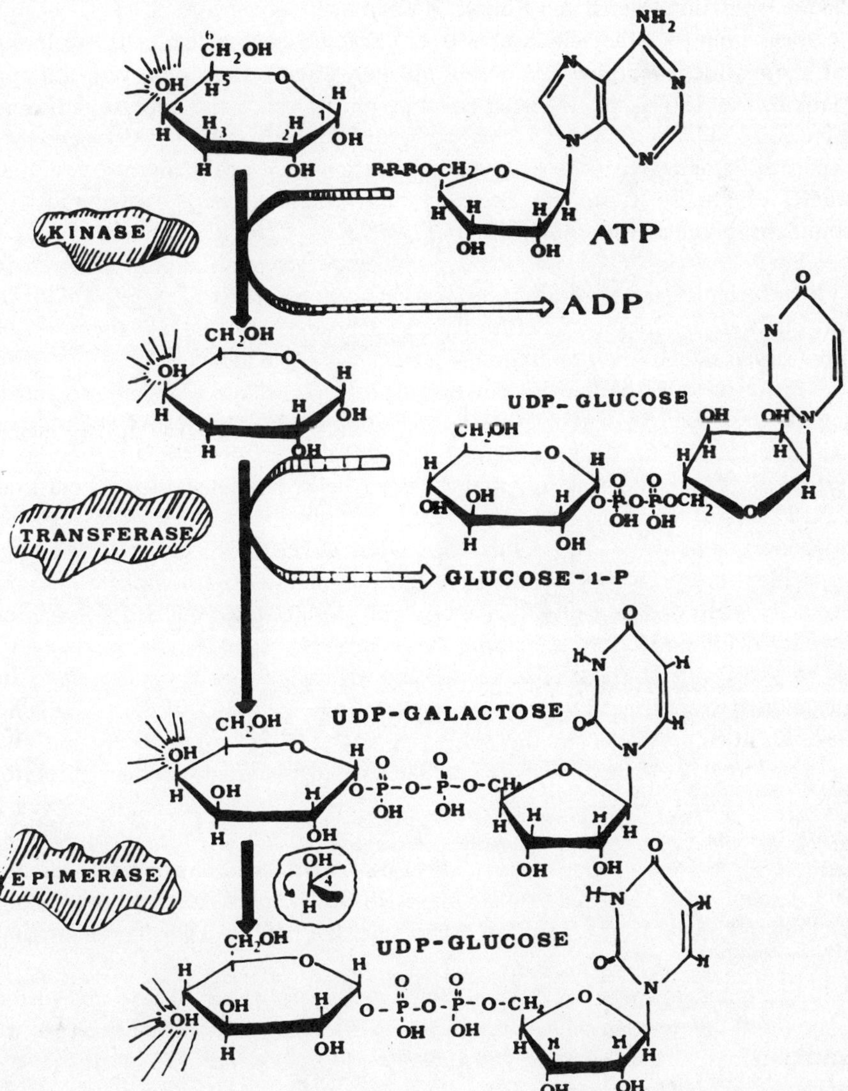

Fig. 1. A diagram of the Leloir pathway. The main metabolic pathway for galactose metabolism for both man and E. coli. One of the major effects of this pathway is the conversion of the galactose moiety to a glucose moiety by a shift in the orientation of the number 4 hydroxyl group.

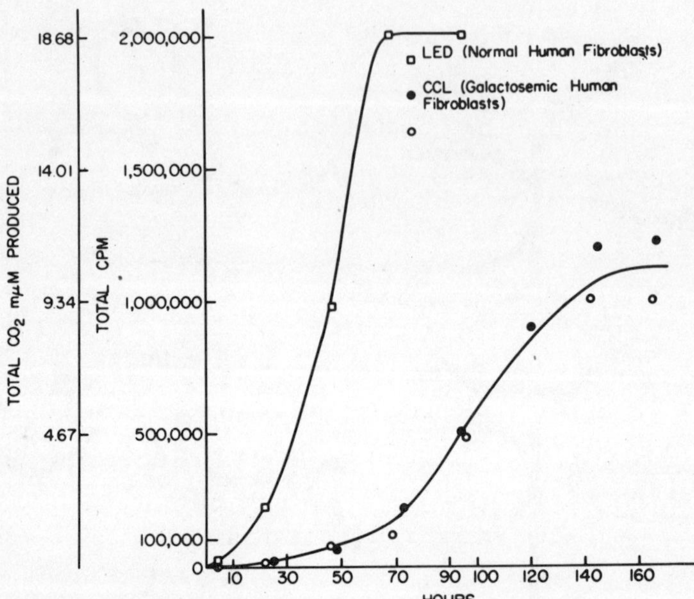

Fig. 2. A screening procedure to detect viral induced changes in galactose metabolism was devised using the production of $^{14}CO_2$ from galactose-1- ^{14}C (Gal-1- ^{14}C) as a measure of the ability of the cells to utilize galactose.

Production of $^{14}CO_2$ was monitored for normal human fibroblast (strain Led-130) and galactosemic fibroblasts, (strain CCL-72). Cultures containing approximately 1×10^5 cells per flask in the logarithmic phase of growth were rinsed twice with Tris buffered saline and inoculated with 1.0 ml of fluid containing bacteriophage (1.4×10^{10} pfu/ml) in Earle's balanced salt solution without sodium bicarbonate. After 45 minutes at room temperature with frequent gentle rocking, the cultures were rinsed twice with Tris buffered saline. Fresh, hexose-free, medium MAB 87/3 containing penicillin and 10% dialyzed fetal calf serum was added. Galactose-1- ^{14}C (1.05×10^7 CPM; 98 muM) was added to each flask and silicone stoppers were inserted. The flasks were then incubated at 34° C. $^{14}CO_2$ was monitored.

Each sample for $^{14}CO_2$ monitoring was obtained by piercing the stopper with a 19 1/2 guage needle and removing 10 cc. of gas anaerobically in a glass syringe whose plunger had been coated with paraffin oil to prevent leaking. Then 10 cc of air was passed through a 0.22μ millipore filter and added to the flask to maintain normal pressure. The collected gas was passed through a column CO_2 trap containing 0.48 ml of phenethylamine and 1 gm of chromosorb W. This system was capable of trapping 99.0% of the CO_2 produced and had a counting efficiency of 65% in 10 ml of toluene-Fluor. A second trap consisted of a bubbler with 5 ml of phenethylamine, and was used primarily to monitor the flow rate of gas which was not allowed to exceed 10 cc/minute (Stuart and Williams, 1967).

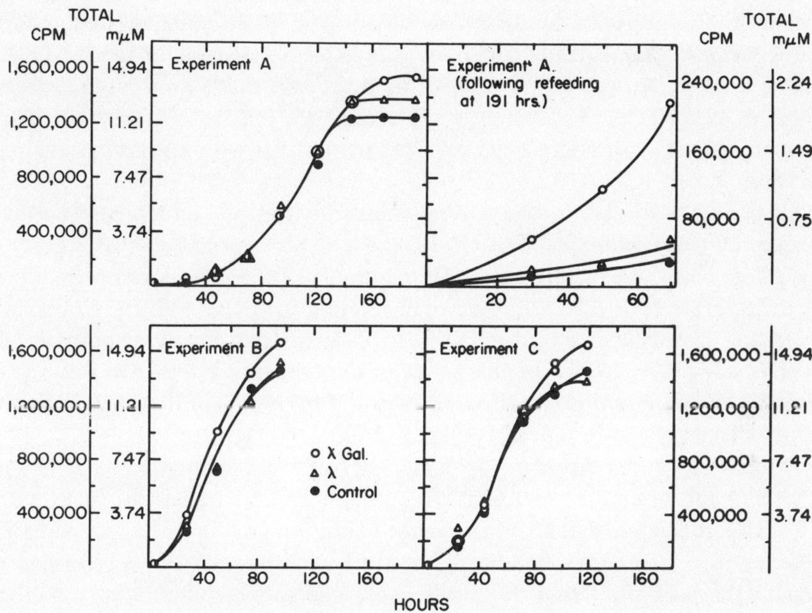

Fig. 3. To test the effect of λ and λpgal on human galactosemic fibroblasts, replicate flasks of CCL-72 cells were inoculated with λ, λ pgal or were not inoculated with virus. After Gal-1- ^{14}C was added, the flasks were sealed and monitored for $^{14}CO_2$ production serially over time. Three representative experiments are illustrated and in all cases the cells exposed to λpgal produced more $^{14}CO_2$ than either those exposed to λ or non-infected cells. This increased $^{14}CO_2$ production has ranged from 150,000 to 300,000 CPM above the control cells. The fourth graph in the upper right corner shows the effect (after 191 hours of incubation) of re-feeding with fresh medium containing Gal-1- ^{14}C in the same concentration as used initially. All flasks of a given experiment were replicate cultures of cells of the same passage and maintained under identical conditions prior to use.

total amount of CO_2 produced by cells in culture is dependent on the extracellular concentration of galactose. Normal cells can produce CO_2 more efficiently from galactose which is at a high concentration in the media, but as the concentration of galactose is reduced the normal and galactosemic cell types appear more and more similar. Further investigations on additional lines of galactosemic cells without detectable gal-transferase and on galactokinase deficient human fibroblasts suggest that the alternate metabolic pathways require the synthesis of galactose-1-phosphate prior to further metabolism (Friedman *et al.*, 1975).

Although the conversion of radioactive galactose to radioactive carbon dioxide looked encouraging, it was important to demonstrate directly that transferase enzyme was formed after infection with λpgal virus. At the time these experiments were undertaken there were very few assays that were sensitive enough to work with small numbers of cells grown in culture. The most sensitive assay had been developed by Russell and Mars (1967). By utilizing some modifications, including the substitution of thin layer chromatography for the DEAE cellulose used by Russell (1968, 1969) the assay was made even more sensitive (Merril *et al.*, 1971). With this assay it was possible to observe transferase enzymatic activity in galactosemic cells infected either with λpgal virus or λ pgal virus DNA.

All viruses used in this study were extensively purified by repeated banding in cesium chloride gradients. The viruses showed no enzymatic activity prior to being utilized in these experiments. All the original experiments were conducted on galactosemic cells CCL-72 from The American Type Culture Collection. Transferase activity ranged from 5.9% to 29% of that found in normal cells (Merril *et al.*, 1971, 1974). In this series of experiments a λpgal ($K^+T^-E^+$) virus containing an amber point mutation in the transferase gene (T^-) was utilized as a control. This virus was produced for us by Dr. Max Gottesman.

The fact that the enzyme appeared when the cell was infected by λpgal ($K^+ T^+ E^+$) and not by an identical virus with a mutant transferase gene, λpgal ($K^+ T^- E^+$), indicated that the appearance of the enzyme must be related to the transferase (T^+) gene. A number of control experiments were performed to eliminate the possibility that the results were due to a contaminating organism. No transferase activity was observed when the infection was carried out in the presence of cyclohexamide, implying that the enzyme was synthesized by the eucaryotic translation mechanisms. Penicillin, streptomyocin, and mycostatin, however, did not affect the production of the enzyme in this series of experiments. No bacterial, mycoplasmal or fungal contaminants were grown from the human cell cultures utilized in these experiments (See appendix A).

In 1973, Dr. B. Rolfe (Australian National University) visited our laboratory and we obtained three new galactosemic cell lines (GM52, GM53, and GM54) from the Institute for Medical Research (Genetic Mutant Cell Culture Repository, Camden, New Jersey). Infection with λpgal ($K^+ T^- E^+$) gave transferase activity in all of these cell lines, indicating that the original results were not unique to the CCL-72 line (See appendix B). Twenty-eight independent experiments were performed and 68% of these experiments yielded transferase activity after infection with the T^+ virus. It became apparent during the course of these and subsequent experiments that there was a large variation from experiment to experiment in the amount of detectable enzymatic activity found in the galactosemic fibroblasts following infection with the procaryotic virus. In the above series, activities ranged from 0.2% to 35% of the activity found in normal human cells with an average activity of 1% of normal. In this series of experiments no detectable activity was obtained in uninfected cells or cells infected with the λpgal ($K^+ T^- E^+$) virus.

An additional control performed during the course of this experiment included the deliberate contamination of the λpgal ($K^+T^-E^+$) virus with a large excess of transferase enzyme from gal ($K^+T^+E^+$) *E. coli* to test whether the virus could passively contaminate the cells with enzyme. Enough enzyme was added to give 128,000 counts per minute of UDP-gal formed in 25 minutes during our normal assay procedure. On assaying these cells, however, no counts above normal background (45 counts per minute) were observed (Merril, 1974).

In an attempt to characterize the enzyme formed in phage infected human galactosemic cells, a column chromatography system was developed to separate the bacterial from the mammalian enzyme. It was found that human enzyme could be separated from the bacterial by passage of the crude cell lysate through a DEAE-cellulose column with a NaCl gradient (Fig. 4). When the material eluted from this column was separated into two groups of aliquots, one of which was preheated at 40°C. for twenty minutes, the mammalian enzyme could be further differentiated from bacterial enzyme. Human enzyme activity was destroyed while the bacterial enzyme seemed to resist such treatment. On two occasions crude cell lysate from human galactosemic cells was passed through these columns and a small, heat stable peak was observed at approximately the position of the bacterial enzyme. These experiments were difficult due to the great variability in enzyme formed from one experiment to the next and the small amount of enzyme which was usually obtained.

One of the largest problems in these endeavors was our inability to develop a selective system for cells which were infected and were making the bacterial enzyme. Experiments have shown that galactosemic cells are sensitive to high galactose concentrations but this sensitivity is dependent on cell density. The higher the density of galactosemic cells in such media, the greater the ability of the media to kill the cells (Yarkin, 1974). This is the opposite of what one would like in order to select for transformed single cells. Without the ability to select cells that produce the bacterial enzyme it is difficult to enrich or increase the amount of enzyme made per flask.

Another way of demonstrating a direct phage effect in eucaryotic cells utilizes the methods of DNA-RNA hybridization. To study transcription in human fibroblasts, λCI857•S7 was utilized as the infecting phage, (Geier and Merril, 1972). A correlation was found between the amount of virus in each infection and the percentage of λ specific RNA produced. This value reached a plateau (0.2% of the total labeled cellular RNA) approximately 4 days after infection. In one case cells monitored for 41 days following infection continued to produce lambda specific RNA. These experiments were performed in the presence of streptomyocin, penicillin, and mycostatin (50 micrograms per ml, 100 units per ml, and 50 units per ml, respectively). Furthermore, any contaminant would have had to be either intercellular or firmly attached to the cell

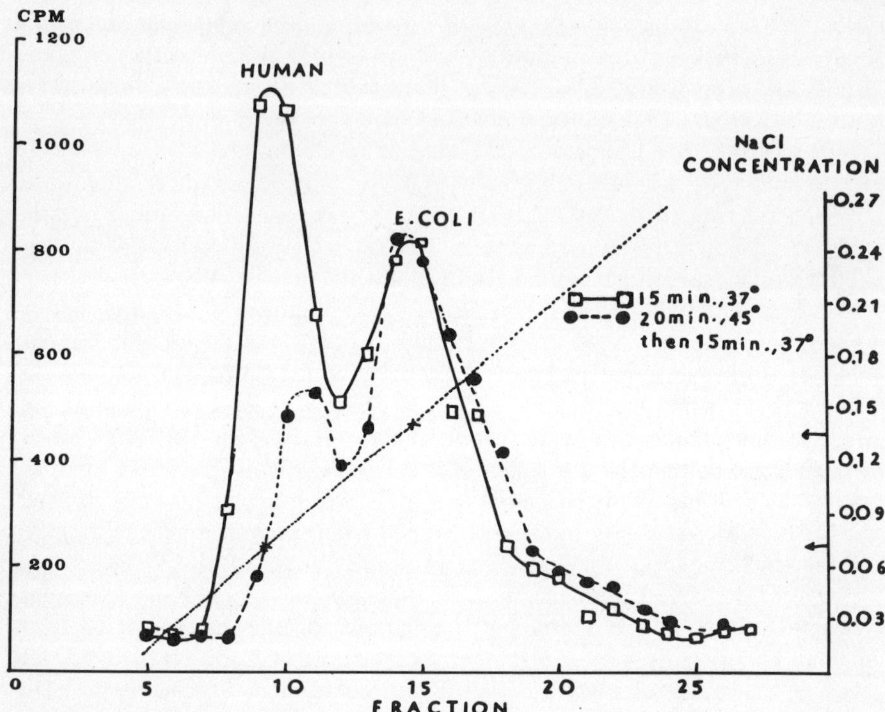

Fig. 4. An ion exchange chromatography method of separating human and bacterial α-D-galactose-1-phosphate uridyl transferase enzymes. The columns utilized DEAE cellulose (DE-52) in a 1 ml column, washed with 0.01M glycine buffer (pH 8.7) containing 10^{-5} M EDTA and 10^{-3}M 2-mercaptoethanol. Parts of 1.2 mg of crude E. coli lysate and 0.6 mg of human fibroblast lysate were mixed together and added to the column in 10 ml of 0.01M glycine buffer (pH 8.7). A 30 ml NaCl linear gradient (0 to 0.3M NaCl) in the glycine buffer was used to elute the enzymes. 1 ml fractions were collected and assayed for enzyme activity. It has been possible to utilize columns with DE-52 column as small as 0.01 ml. Separation of enzymes in a column this small utilizes a step-gradient (0.03M NaCl) with an elution of 0.03 ml/step.

surface; for in an experiment which was deliberately contaminated with an *E. coli* lysogen, the lysogen was removed by normal rinsing procedures employed. No growth was observed by monitoring these experiments with tryptone plates. Furthermore, any contaminant would have to have little homology of its RNA with *E. coli* since hybridization with *E. coli* DNA filters failed to show any significant amount of RNA hybridization.

Currently experiments are being undertaken to define the initial events after mammalian cells are infected with bacterial viruses. Such studies may make the production of bacterial enzymes in these systems a more reproducible phenomenon. Studies utilizing λ phage labeled with tritiated thymidine in infections of mammalian cells in culture showed a maximal adsorption 30-40 hours after infection, with multiplicities of phage of 1.5 x 10^5 (Fig. 5). This adsorption rate

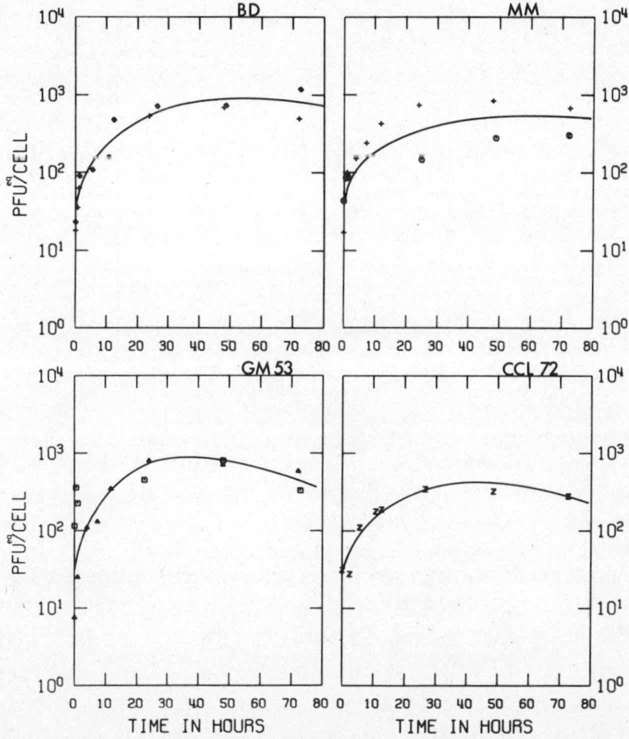

Fig. 5. Adsorption curves obtained by infecting human fibroblasts with radioactive λCI857 virus. Cell lines GM53 and CCL-72 are from individuals with galactosemia. Multiplicity of infection was 7 x 10^4 pfu/cell. Infections were performed in Earle's balanced salt solution. One hour after infection, Earle's MEM with 10% fetal calf serum was added (defined as time 0 on the graphs.) Cells were assayed by decanting the flasks, rinsing 4 times with 5 ml Earle's MEM (with no serum) and twice with phosphate buffer saline containing 0.1 M $MgSO_4$. The cells were scraped from the flask and trapped in a glass fiber filter (type A, Gelman). The filters were dried, placed in a toluene scintillation cocktail, and counted.

appeared relatively constant and independent of the cell line used. As the adsorption of radioactive phage increased, the number of detectable phage plaque forming units (pfu) decreased (Fig. 6). This observation is compatible with an uncoating of the viral DNA.

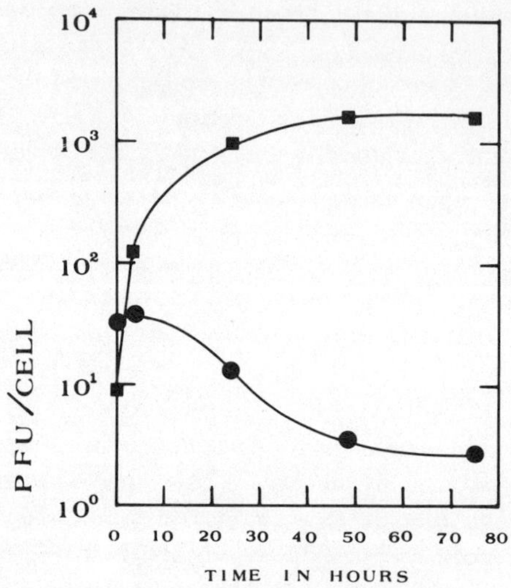

Fig. 6. Adsorption and plaque forming ability of adsorbed phage in mammalian cells fibroblasts. The cell line used in this experiment was the 3T3 mouse fibroblast cells. Experiments were performed as described in Fig. 5, with the additional step of titering (for plaque forming ability) a small aliquot, from the washed cells prior to filtering, against E. coli.

We are now studying the molecular fate of the adsorbed DNA. Recent findings have shown that DNA from phage T7 grown on *E. coli* B is extensively degraded after entry into human fibroblasts (Kao *et al.*, 1973). It has been found, however, that T7 DNA taken up by golden Syrian hamster embryo cells is not totally degraded. It is localized in the nuclei. Sedimentation analysis as well as DNA-DNA hybridization studies indicates that the DNA is high molecular weight T7 DNA (Schechtman *et al.*, 1973). Whether the bacterial host used to grow the phage, the type or state of the recipient cells, or other experimental variables affected the fate of the T7 DNA will have to be determined in future experiments.

Studies on the interaction of phage in whole animal systems have also been initiated. It was found that the distribution of bacterial viruses was specific regardless of the injection route. The virus titers observed in most organs studied decreased exponentially; however, the spleen retained relatively high titers of phage (10^5 per ml) for periods of up to seven days (Geier *et al.*, 1973). Since the bacterial viruses observed in the spleen were still capable of forming plaques in the *E. coli* it was postulated that these phages were trapped by a non-phagocytic type of process.

It may be possible to detect bacterial gene function following infection with bacterial viruses by using special diets for mice. We have developed two diets which may be of some assistance in this endeavor. The first diet is complete except for the lack of the amino acid leucine (Trigg *et al.*, 1975). Mice on this diet generally survive for only 60 days. In our laboratory we have isolated a lambda virus carrying the leucine operon (Klingmuller and Krell, 1973) and are currently testing this virus in mice on the leucine diet.

The other diet that has been developed utilizes a high concentration of galactose. Mice on this diet die within 15 days. We are also currently attempting to see if mice infected with λpgal can survive longer than mice infected with lambda or non-infected mice on the high galactose diet. Rolfe and Richardson in Australia have administered λpgal virus to the grey kangaroo which has little transferase activity (Stephens *et al.*, 1974). They have found evidence of bacterial enzyme seven days after infection with λpgal virus (Rolfe and Richardson, 1975).

These studies on bacterial virus interaction with mammalian systems are in a very early stage. In view of the long and continued relationship between man and numerous bacteria, it is difficult to imagine that there would not be at least some viruses capable of replicating in either host. Such a virus could serve an important link in probing both procaryotic and eucaryotic systems. It may be that much of the work that has been performed with lambda virus has been done under relatively forced conditions, and some other virus, more adept to growth in alternate hosts, might give better results. The search for a virus with these capabilities will be difficult. It is much like the proverbial search for the needle in the haystack and there are at present few clues to identify such a virus.

One intriguing candidate might be the polio virus. In 1942 a large amount of polio virus was reported in the sewers of Stockholm (Kling *et al.*, 1942). The investigators in this study concluded that the amount of virus found required no fewer than 100,000 humans infected with polio; however, there were only three known cases at the time in the district. One viable model was that the polio virus multiplied after leaving the gastrointestinal tract of the infected individuals. Most research on the possibility of polio replication in sewage material has concentrated on protozoa (Toomey *et al.*, 1948; Young *et al.*, 1948; Brutseart *et al.*,

1948). None of the above studies demonstrated any ability of the polio virus to replicate in protozoa or some of the more easily cultured bacteria. It should be pointed out, however, that the main organisms in the human gastrointestinal tract are not *E. coli* but the various anaerobes, such as the bacteriodies. No reports of attempts to replicate polio in these organisms under anerobic conditions have been published.

It has been shown that polio virus RNA can stimulate amino acid incorporation by an *E. coli*-cell-free system (Warner *et al.*, 1963; Rekosh *et al.*, 1969). Polio virus RNA is as efficient as f2 bacteriophage RNA in stimulating amino acid incorporation in *E. coli*-cell-free systems. Tryptic peptides formed in this *E. coli*-cell-free system correspond to tryptic peptides from authentic polio virus protein; however, some tryptic peptides produced by digestion of authentic polio protein are not found in the *in vitro* product. The polio RNA is rapidly degraded in the cell-free system, and this degradation is believed to be responsible, to some extent, for the heterogeneity of the product (Rekosh *et al.*, 1970). There have been some reports of replication of mammalian viruses within bacteria; however, each of these has been a single isolated report with no confirmation (Abel and Trautner, 1964; Bayreuther and Romig, 1964; Babbar, 1968).

The search for bacterial viruses which can live in both mammalian and bacterial hosts should prove to be interesting both from a medical and molecular biology point of view. The results from such a search might prove even more important than the possible therapeutic utilization of bacterial viruses for the treatment of specific genetic diseases. The current use of bacteria viruses as vehicles for eucaryotic genetic material (formed by *in vitro* recombination techniques) further increases the importance of determining the potential of these viruses to interact directly with mammalian systems.

ACKNOWLEDGEMENT

The author would like to thank Mrs. Mary Railsback for assistance in the preparation of this manuscript.

REFERENCES

Abel, P., and Trautner, T. A. (1964). Formation of an animal virus within a bacterium, *Z. Vererbungal 95*, 66-72.

Babbar, O. P. (1968). Production of mature virus particles by *Escherichia coli* spheroplasts infected with lipid treated RNA Ranikhet disease virus and with Vaccinia and Ranikhet disease virus. *Acta Virologia 12*, 497-506.

Bayreuther, K. E., and Romig, W. R. (1964). Polymer virus: production in *Bacillus subtilis*. *Science 146*, 778-799.

Brutseart, P., Jungeblut, C. W., and Knox A. (1948). Attempts to propagate murine poliomyelitis virus on various intestinal bacteria and protozoa. *Proc. Soc. Exp. Bio. Med. 61,* 165-269.

Centifanto, Y. M. (1968). Antiviral agent from λ-infected *E. coli* K12. 1. Isolation. *Applied Microbiol. 16,* 827-834.

D'Herelle, F. (1917). Sur un microbe invisible antagoniste des bacilles dysenteriques. *C. R. Acad. Sci. (Paris) 165,* 373.

D'Herelle, F. (1918). Sur la role du microbe filtrant bacteriophage dans la dysenterie bacillaire. *C. R. Acad. Sci. (Paris) 167,* 970.

Friedman, T. B., Yarkin, R. J., and Merril, C. R. (1975). Galactose and glucose metabolism in galactokinase deficient, galactose-1-P-uridyl transferase deficient and normal human fibroblasts. *J. Cell Physiol. 85,* 569-578.

Geier, M. R., and Merril, C. R. (1972). Lambda phage transcription in human fibroblasts. *Virology 47,* 638-643.

Geier, M. R., Trigg, M. E., and Merril, C. R. (1973). The fate of bacteriophage lambda in non-immune germfree mice. *Nature 246,* 221-223.

Kao, P. C., Regan, J. D., and Volkin, E. (1973). Fate of homologous and heterologous DNAs after incorporation into human skin fibroblasts. *Biochim. Biophys. Acta 324,* 1-13.

Kleinschmidt, W. J., Douthart, R. J., and Murphy, E. B. (1970). Interferon production by T4 coliphage. *Nature 228.* 27-30.

Kling, C., Olin, G., Fahreus, J., and Norlin, G. (1972). Sewage as a carrier and disseminator of poliomyelitis virus. *Acta Med. Scand. 122,* 217-263.

Klingmuller, W., Krell, K., and Shimada, K. (1973). Genetic analysis of bacteriophage lambda strains which transduce genes for leucine biosynthesis. *Mole. Gen. Genet. 126,* 1-6.

Mankiewicz, E. (1965). Bacteriophages that lyse mycobacteria and corynobacteria and show cytopathogenic effect on tissue cultures of renal cells of cercopithecus aethlops: A preliminary communication. *J. Can. Med. Assoc. 92,* 31.

Meek, E. S., and Takahashi, M. (1968). Differential inhibition by phagicin of DNA synthesis in cells infected with *Vaccinia. Nature 220,* 822.

Merril, C. R., Geier, M. R., and Petricciani, J. C. (1971). Bacterial virus gene expression in human cells. *Nature 233,* 398-400.

Merril, C. R., Friedman, T. B., Attallah, A. F. M., Geier, M. R., Krell, K., and Yarkin, R. (1973). Isolation of bacteriophages from commercial sera. *In Vitro 8,* 91.

Merril, C. R., Geier, M. R.,and Trigg, M. E. (1974). Transduction in mammalian cells. *In Birth Defects* (A. G. Motulsky, and W. Lentz, eds.), pp. 81-91. Excerpta Medica, Amsterdam, Holland.

Merril, C. R. (1974). Bacteriophage interactions with higher organisms. *Trans. N.Y. Acad. Sci. 36,* 265-272.

Petricciani, J. C., Binder, M. K., Merril, C. R., and Geier, M. R. (1972). Galactose utilization in galactosemia. *Science 175,* 1368-1370.

Rekosh, D., Lodish, H. F., and Baltimore, D. (1969). Translation of polio virus RNA by an *E. coli* cell free system. *Cold Spr. Harb. Symp. Quant. Biol. 34,* 747-754.

Rekosh, D., Lodish, H. F., and Baltimore, D. (1970). Protein Synthesis in *Escherichia coli.* Extracts programmed by polio polio virus RNA. *J. Mol. Biol. 54.* 327-340.

Rolfe, B., and Richardson, B. J. (1975), personal communication.

Russell, J. D. (1969). Improved transferase assay for cultured fibroblasts. *Biochem. Genetics 1,* 301-303.

Russell, J. D. (1969). Variation in UDP Glu: Alpha-D-Gal-1-P Uridyl transferase activity during growth of cultured fibroblasts. In *Galactosemis* (D.Y.Y. Hsia, ed.) pp. 204-212. C. C. Thomas, Springfield, IL.

Russell, J. D., and DeMars, R. (1967). UDP glucose:α-D-galactose-1-phosphate uridyl transferase activity in cultured human fiborblasts. *Biochem. Genetics 1,* 11-24.

Schechtman, L. M., Leavitt, J. C., Ts'O, P., Borenfruend, E., and Bendich, A. (1973). Uptake of T7 phage and phage DNA by hamster cells in culture. *Biophys. Soc. Abstr. 326a.*

Stephens, T., Irvine, S., Mutton, P., Gupta, J.D., and Harley, J.D. (1974). Deficiency of two enzymes of galactose metabolism in kangaroos. *Nature 248,* 524-525.

Stuart, S. C., and Williams, G. R. (1967). $^{14}CO_2$ Measurements during mitochondrial respiration. In *Methods in Enzymology 10,* (R. W. Esterbrook and M. E. Pullman, eds.), pp. 650-655. Academic Press, N.Y.

Toomey, J. A. Takacs, W. S., and Schaefter, M. (1948). Attempts to affect amoeba proteins with poliomyelitis. *Am. J. Dis. Child. 75,* 11-18.

Trigg, M. E., Geier, M. R., LaPolla, R. J., Kamerow, H. N., and Merril, C. R. (1975). Addition of leucine precursors to the diet of leucine-starved mice. *Am. J. Clin. Natr. 28,* 947-949.

Warner, J. R., Madden, M. J., and Darnell, J. E. (1963). The interaction of polio virus RNA with *E. coli* ribosomes. *Virology 19,* 393-399.

Young, V. M. Felsenfeld, O., and Byrd, C. J. (1949). Behavior of virus of poliomyelitis in a culture of *Endamoeba histolytica, Am. J. Clin. Pathol. 19,* 1135-1138.

APPENDIX A: Phage Gene Directed Enzyme Activity in Genetically Defective Human Cells

Jurgen Horst, Fridrich Kluge, Konrat Beyreuther and Wolfgang Gerok

One of the best known regulatory mechanisms of microorganisms concerns the *lac* operon of *Escherichia coli.* The introduction of *lac* operon DNA into mammalian cells and its expression there might therefore be useful in contributing to our understanding of regulatory mechanisms in eucaryotic cells.

For gene transfer experiments in human cells we used the specialized transducing phage λ plac carrying a complete *z* gene, the gene for *E. coli* β-galactosidase. As recipient cells, human skin fibroblasts were grown from a patient with generalized gangliosidosis with a severe difficiency of β-galactosidase activity. The deficient human fibroblasts were incubated with the bacterial virus λplac or with λplac DNA. As control, uninfected deficient fibroblasts were used as well as fibroblasts treated with wild type λphage.

We have detected a significant overall increase of the β-galactosidase activity in about 15% of the λplac phage treated and in about 25% of the λplac DNA treated cells. Induced enzyme activity varied from 0.8 to over 300 nmoles of 4-methylumbelliferone liberated in 60 minutes/mg protein at 37°C (Horst *et al.,* 1975).

Three criteria make it likely that the induced enzyme activity is directed by the offered procaryotic genetic material. (i) The induced enzyme activity can be measured at pH 6.8, the pH optimum of *E. coli* β-galactosidase (Wallenfels and Weil, 1974), and not at pH 4.0, where *E. coli* β-galactosidase is inactive. (ii) The work of Melchers and Messer (1970) demonstrated that *E. coli* β-galactosidase can be protected against heat in activation by anti-*E. coli* β-galactosidase. Investigation of the detected β-galactosidase activity in cells after treatment with phage λplac or λplac DNA revealed the *E. coli* origin of the β-galactosidase by means of this immunochemical behavior. (iii) As shown in Fig. 1, normal fibroblasts β-galactosidase and *E. coli* β-galactosidase can be separated by ion exchange chromatography; theλplac induced β-galactosidase activity in the deficient fibroblasts co-chromatographed with *E-coli β-galactosidase.*

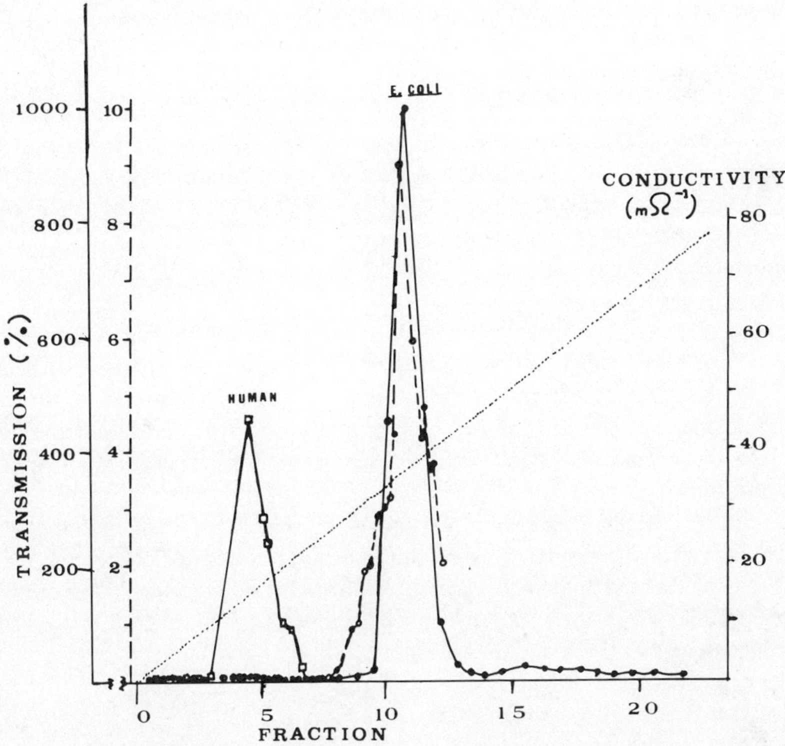

Fig. 1 A composite graph of elution profiles of β-galactosidase activity from ion exchange chromatography on DEAE-cellulose. Columns were eluted from 0-0.7 M in 0.01 M sodium phosphate buffer, pH 6.8. The E. coli enzyme activity eluted at a conductivity of 37 m mho (o—o). Human enzyme eluted at 15 m mho (□—□). Enzyme obtained in λplac DNA infected G_{M1} fibroblasts eluted at 36.6 m mho (o---o).

REFERENCES

Horst, J., Kluge, F., Beyreuther, K., and Gerok, W. (1975). Gene transfer to human cells: Transducing phage λplac gene expression in G_{M1}-gangliosidosis fibrobalsts. *Proc. Nat. Acad. Sci. (U.S.A.) 72,* 3531-3535.

Melchers, F., and Messer, W. (1970). Enhanced stability against heat denaturation of *E. coli* wild type and mutant β-galactosidase in the presence of specific antibodies. *Biochem. Biophys. Res. Commun. 40,* 570-575.

Wallenfels, K., and Weil, R. (1974). β-galactosidase. In *The Enzymes 7,* (P.D. Boyer, ed.), pp. 617-663. Academic Press, N.Y.

APPENDIX B: Phage - Mediated Transgenosis in Human Galactosemic Fibroblasts

Barry G. Rolfe

Some of the experiments described here were carried out independently while others were conducted as a group effort during a visit to the National Institute of Mental Health in the laboratory of Dr. Carl Merril. Three galactosemic cell lines were used. Variable levels of galactosyl transferase activity (following the infection with the phage $\lambda pgal^+$) were found in 68% of the experiments. This variation is reminiscent of the early work on transformation in *Pneumococcus* and the much studied phenomenon of "cell competence" (Hotchkiss, (1954).

Our results suggested that there were at least two key points in the experimental design which might greatly influence the success or failure of demonstrating the bacterial transferase activity in human cells. The first has to do with the growth phase of the cells at the time of phage infection and the second, the optimum time to harvest $\lambda pgal^+$ infected cultures and to assay for the bacterial transferase activity.

In studies described elsewhere (Merril *et al.*, 1974), it was concluded that the most reproducible results were obtained by infecting rapidly growing (mid-logarithmic phase) cultures with high titres of transducing virus. It should be pointed out that all of these experiments described above were done with media containing glucose as the main hexose source.

The question of whether the introduced bacterial genes can function effectively *in vivo* within the mammalian cell was answered by experiments using replica flasks of the galactosemic line GM52 (Merril, this volume). Radioactive galactose-1- ^{14}C was used as the *sole* hexose source. It had been found earlier that normal human fibroblasts could convert about two to three times as much galactose to CO_2 as the galactosemic cell cultures (Merril *et al.*, 1974). The GM52 cell cultures in galactose-1-^{14}C media were infected with either phage $\lambda pgal^+$, λ^+or with phage $\lambda pgal^-$

In five separate experiments it was found that after 5 to 6 days those galactosemic cells exposed to phage $\lambda pgal^+$ had produced between 11% and 14% more $^{14}CO_2$ from galactose-1-^{14}C than the non-infected, λ^+ infected or phage $\lambda pgal^-$ infected cell controls.

The optimum length of incubation of $\lambda pgal^+$ infected cultures before harvesting was determined by setting up a series of replica flasks of GM52 cells containing either galactose or radioactive galactose as the hexose source. These cultures were infected with either phage $\lambda pgal^+$ or $\lambda pgal^-$. The galactosyl transferase activity of bacterial origin was most reproducibly detected in those human cell cultures which had been incubated for about 6 days, the time of maximum enhanced galactose utilization (Fig. 1). Enzyme assays carried out on cell cultures at times before maximum CO_2 formation often failed to show any

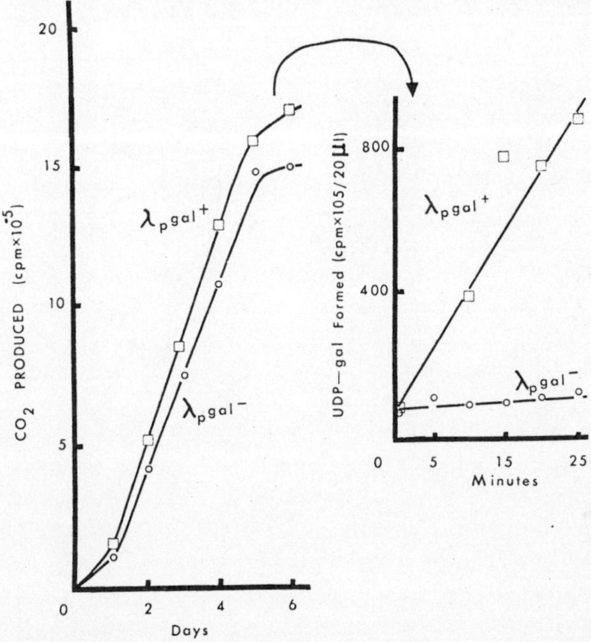

Fig. 1. Transferase activity detected at the time of galactose metabolic enhancement in galactosemic fibroblasts. The culture conditions and the monitoring of $^{14}CO_2$ were the same as those described by Merril (this volume). Different cultures were assayed after the 4th, 5th, and 6th days of incubation for the bacterial transferase activity.

transferase activity. However, the situation remains complicated by the fact that enzymatic activity could not be measured in all cultures exhibiting this enhanced $^{14}CO_2$ yield as a result of λpgal⁺ infection. Various procedures for lysing cells (such as detergents, sonication, freezing and thawing) have been used but there is no evidence that any particular procedure has an advantage over the others.

At present, it seems that some factor(s) in the human cells may cause a lability of the bacterial transferase activity upon or during the harvesting of the λpgal⁺ cultures. Experiments are in progress to see if this inactivation is due to host protease systems.

REFERENCES

Hotchkiss, R. D. (1954). Cyclical behavior in *Pneumococcal* growth and transformability occasioned by environmental changes. *Proc. Nat. Acad. Sci. (U.S.A.). 40,* 49-55.

Merril, C. R., Grier, M. R., and Rolfe, B. G. (1974). Characteristics of bacterial gene expression in human fibroblasts. In *The Eucaryote Chromosome* (W. J. Peacock and R. D. Brock, eds.), Australian National University Press.

Gene Transfer in *Drosophila melanogaster*

Allen S. Fox

Central to the problems of genetic molecular modification of eucaryotes are the prospects for the successful accomplishment of gene transfer. Such prospects have existed since the discovery of bacterial transformation about thirty years ago, but it is only recently that systems and techniques have been developed which permit the critical demonstration of genetic transfer in eucaryotes. The purpose of this paper is to review the results obtained with one such system, and to draw inferences from these results with respect to prospects and applications in other organisms.

EXPERIMENTAL STRATEGIES

The experimental strategies attempted or envisaged for gene transfer in eucaryotes may be classified in a two-dimensional matrix, depending on what is treated (whole organisms or cells in culture) and how the genetic information to be transferred is administered.

Successful transfer of genetic information into a whole organism would result in the introduction of specific genes into somatic cells (evidenced by a phenotypic change in the daughter cell) or into germinal cells (evidenced by the permanent genetic transformation of their progeny). Genetic transfer into cultured cells would be signaled by a specific phenotypic change in some of the treated cells or their progeny, and by transmission and expression of their genes in their cell lineages. Phenotypic alteration of whole organisms might be achieved by the introduction of such transformed cell lineages into somatic tissues, but transmitted genetic transformation could be achieved only if the introduced lineages could be established in the germ line.

Each of these approaches has its peculiar advantages and disadvantages. The major advantage of the whole-organism approach is that it affords an opportunity for exhaustive genetic analysis, provided the proper organism is chosen. The major disadvantage is that events at the cellular level can be studied only with difficulty. The major advantage of the cell-culture approach is the opportunity it offers for analysis of cellular events and for clonal analysis, while its greatest disadvantage stems from the unfortunate fact that techniques for genetic analysis of cells in culture are currently severely limited in scope. Taken together, the two approaches complement each other in a manner which could result in increased understanding of the mechanisms by which transferred genes may be established in eucaryotic cells.

The most direct approach to introduction of genetic information is afforded by the model of bacterial transformation, utilizing naked DNA. The work reviewed here is an example of this approach. A possible difficulty is the susceptibility of such DNA to degradation by nucleases. Protection might be afforded by the use of chromatin (nucleohistone) - which to my knowledge has not been attempted - or by the use of isolated metaphase chromosomes (McBride and Ozer, 1973; Willecke and Ruddle, 1975). Alternatively, the use of viral vectors has been suggested; either in the form of spontaneously produced transducing viruses (Shimada *et al.*, this volume; Merril, this volume), or in the form of plasmids or viruses into which segments of eucaryotic DNA have been inserted by the use of restriction enzymes. Another intriguing possibility is the transfer of genetic information by RNA, provided establishment in the host DNA could be accomplished by reverse transcriptase (Mishra *et al.*, 1975). Finally, transfer might be accomplished by cell fusion (Schwartz *et al.*, 1971).

Each of these methods of administration will find its preferred applications depending on the system. A question of fundamental importance is whether the nature of the association of the introduced genetic information with the host genome will be the same in all cases.

DROSOPHILA AS AN EXPERIMENTAL SYSTEM

The strategy of gene transfer by treatment of the whole organism was adopted for the work reviewed in this paper. Several considerations motivated the choice of this strategy: 1) The immediate object of gene transfer is to achieve a phenotypic alteration of whole individuals by the introduction of exogenous genetic information. Regarding "gene therapy", the objective is to achieve a phenotypic cure of individuals suffering from genetic defects (Friedmann and Roblin, 1972). 2) The long-range object is to achieve permanent transmission of the introduced information by genetic substitution in the germ line. In terms of gene therapy this would constitute permanent genetic cure of whole individuals in subsequent generations. 3) Most importantly, it seemed to us that neither of these objectives could be accomplished without a thorough understanding of the genetic mechanisms involved demanding the most sophisticated tools of genetic analysis.

Such genetic tools are provided by *Drosophila melanogaster,* including refined cytogenetics of the polytene salivary gland chromosomes and favorable developmental features. Following fertilization, the nuclei of the embryo undergo a series of extremely rapid mitoses in a syncytial cytoplasm undivided by cell membranes (Sonnenblick, 1950); only later is the syncytium divided into individual blastoderm and yolk cells. DNA used for treatment at this time needs only to penetrate the egg membrane in order to reach chromosomes at any stage of the mitotic cycle, including that of rapid DNA synthesis. Normal eggs containing embryos at the preblastoderm stage of development are surrounded by two envelopes, an outer chorion and an inner vitelline membrane, which render them impermeable even to small molecules. If, however, females are forced into rapid oviposition, they lay eggs with defective vitelline membranes; removal of the chorion with sodium hypochlorite renders these eggs permeable even to macromolecules. A device known as the "ovitron" makes it possible to collect large numbers of such eggs (Yoon and Fox, 1965). Treatment is performed by simple immersion of the eggs, after chorion removal, in DNA solutions. If the DNA is labeled with ^{32}P or ^{3}H, autoradiography reveals massive label uptake (Yoon and Fox, 1965).

Details of DNA preparation and egg treatment have been given in previous publications (Fox *et al.*, 1975). DNA was prepared from adult flies and included both main band and satellite components. It was double-stranded, with a melting point of 84° C in 0.195 M NaCl and a G + C content of 39.1%. It was disperse in molecular size, with a modal molecular weight of about 10^6 daltons, and was used at a concentration of 0.02 mg/ml.

Since the embryos possessed a few to about a thousand nuclei at the time of treatment, it would be anticipated that pieces of DNA representing a particular gene would at best enter only one or a few nuclei in each embryo, yielding adults which would be mosaic for that gene; that is, they would possess one or a few patches of tissue with the exogenous genetic information surrounded by genetically unaltered tissue. The phenotypic manifestation of such genetic mosaicism would depend on the manner in which the introduced gene produced its effects. If it was autonomous in expression, that is, if the phenotype of a patch of tissue depended on its own genotype and not on that of surrounding tissue, and if the mosaic patch was located in a position where its phenotype was observable, then the fly would exhibit phenotypic mosaicism. When the mosaic patch is not located in a detectable position, its presence would not be observed. If, on the other hand, the introduced gene is non-autonomous in expression, that is, if it produces a diffusible product, the mosaic patch need not be located in tissue manifesting the observed phenotype; diffusion of the product could affect the phenotype even at a distance.

Occasionally such a patch might extend into a gonad; gametes produced by the mosaic sector would possess the introduced gene and would transmit it to the next generation. The proportion of progeny receiving the introduced gene would not be Mendelian, but would depend on the relative size of the gonadal mosaic sector.

Since, in *Drosophila,* germ cells descend from some three to seven nuclei which penetrate into the posterior polar plasm during the preblastoderm period (Sonnenblick, 1950), no necessary correlation between observable phenotypic effects of DNA treatment and transmission to subsequent generations is anticipated. Among treated flies, most phenotypically altered individuals would not be expected to transmit the alteration since it is unlikely that the introduced information would be present both in somatic tissues and in the germ line unless it were introduced very early into those few stem nuclei which give rise to both somatic and polar nuclei. On the other hand, if the introduced information is present in the germ line but not in the soma, phenotypically unaltered flies might produce altered progeny.

It should be noted that if an introduced DNA segment representing a specific gene is integrated into the recipient chromosome (that is, with replacement of the homologeous host segment), as is usually the case in the bacteria transformation (Fox and Allen, 1964), transmission through a gamete would give rise to a whole-body transformant in the next generation. As we shall see, it is in this respect that our results with *Drosophila* differ most markedly from those of bacterial transformation.

In actual practice, in each experiment eggs were divided into at least three different treatment groups. A portion were treated with *allo-DNA,* DNA prepared from flies differing from those being treated with respect to one or more specific genes. Another portion were treated with *iso-DNA,* DNA prepared from flies of the same genotype as those being treated. The third portion was treated with the medium in which the DNA was dissolved (Ringer solution or sucrose). The effects of allo-DNA were evaluated by comparison with the effects of iso-DNA, while comparison of the effects of iso-DNA with those of the non-DNA treatment served to evaluate non-specific effects of DNA. Care was taken to randomize eggs among the various treatments, and a coding system was used which prevented the observer of the treated flies from knowing the treatment to which the eggs were subjected (Fox *et al.*, 1975).

Target loci need to be chosen with care. If the scoring and evaluation of induced mosaicism is the objective, autonomous genes are appropriate but should produce effects which can be clearly detected even in small mosaic sectors. Bristle effects are especially suitable, since each bristle is the product of a single cell (Lees and Waddington, 1942), but genes producing autonomous effects on other traits, including eye color, are also useful. Non-autonomous genes are appropriate where the total yield of induced changes needs to be evaluated since non-autonomy results in a kind of amplification of the effects of small mosaic sectors.It also makes possible detailed genetic analysis, which otherwise is rendered impractical by the necessity to search for fine-grained mosaicism for autonomous traits. Non-autonomous eye-color genes are useful in this regard.

Selective systems, both genetic and biochemical, can be devised for the screening of rarely induced events.

SUMMARY OF EXPERIMENTAL RESULTS

Treatment of young *Drosophila* embryos with allo-DNA induced specific genetic effects not induced by treatment with iso-DNA. The following facts have been established with regard to this phenomenon (Fox *et al*., 1971b).

1. *Eleven loci have been altered by allo-DNA, involving a variety of phenotypic effects both autonomous and non-autonomous. Changes have been induced from recessive allele to dominant, as well as dominant to recessive.*

The data on which these statements are based are given in Fox and Yoon, 1966 and 1970. The genes which have thus far been observed to respond to treatment with allo-DNA are listed in Table 1. In each case iso-DNA has failed to induce similar changes.

TABLE 1*

Genes responding to treatment with allo-DNA.

Gene**	Position on chromosomal map	Direction of induced change
	X-chromosomal genes	
yellow (y)	0.0	$y \rightarrow y^+$
scute (sc)	0.0	$sc \rightarrow sc^+$
white (w)	1.5	$w \rightarrow w^+$
crossveinless (cv)	13.7	$cv \rightarrow cv^+$
singed (sn)	21.0	$sn^3 \rightarrow sn^+$
vermilion (v)	33.0	$v \rightarrow v+$
	3rd-chromosomal genes	
roughoid (ru)	0.0	$ry+ \rightarrow ru$
hairy (h)	26.5	$h+ \rightarrow h$
scarlet (st)	44.0	$st+ \rightarrow st$
curled (cu)	50.0	$cu+ \rightarrow cu$
claret (ca)	100.7	$ca+ \rightarrow ca$

*From Fox *et al*., 1971b.
**Descriptions of these genes are given by Lindsley and Grell (1968).

These genes are located at positions distributed along the lengths of two different chromosomes, and affect a variety of phenotypic traits: eye color (*w, v, st, ca*), bristle color (*y*), bristle shape *(sn)*, bristle number or distribution *(sc, h)*, eye facet arrangement *(ru)*, wing venation *(cv)*, and wing shape *(cu)*. Thus, the effects of allo-DNA appeared to be ubiquitous. A few cases have been encountered in which target genes failed to respond to allo-DNA *(ry^2, cn, bw, mal^{bz}, th)*. There may be a real difference in genic susceptibility, but it seems more probable that these failures were attributable to more trivial explanations. Possibilities include technical deficiencies in the pertinent experiments, inappropriate phenotypic criteria for the detection of mosaicism, developmental blocks to the expression of mosaicism, or differences in competence between stocks. In all cases except *vermilion (v)*, the phenotypic changes observed among allo-DNA treated flies were mosaic. The size of these patches varied, but the majority were very small.

The size distribution and frequency of y^+ or sn^+ patches observed among $y\ sn^3$ flies treated with allo-DNA ($y^+\ sn^+$), iso-DNA ($y\ sn^3$), or Ringer solution are given in Table 2.

TABLE 2

Size distribution of mosaic patches among treated $y\ sn^3$ flies.

Treatment	Patch Size				Total
	1 bristle	2 bristles	3 bristles	4 or more bristles	
Allo-DNA ($y^+\ sn^+$)	114 (0.814)	9 (0.064)	12 (0.086)	5 (0.036)	140 (1.000)
Iso-DNA ($y\ sn^3$)	41 (0.872)	3 (0.064)	2 (0.043)	1 (0.021)	47 (1.000)
Ringer	20 (0.820)	2 (0.087)	0 (0.000)	1 (0.043)	23 (1.000)
Total	175 (0.833)	14 (0.067)	14 (0.067)	7 (0.033)	210 (1.000)

Since each bristle was the product of a single cell, the number of bristles in a patch was an adequate measure of its size. Among flies treated with Ringer solution it was presumed that mosaic patches arose from somatic mutations of y to y^+ or sn^3 to sn^+, that these may have occurred at any time during development, and that they were inherited clonally. Patches only one bristle in size represented mutations occurring late in development and their high frequency resulted from the relatively large cell population present at that time. The data exhibit no significant difference in the distribution of patch size among flies subjected to the three different treatments. The meaning of this observation will be discussed below.

The frequencies of mosaic patches for autonomous loci among flies subjected to the three treatments are summarized in Table 3. The total frequency of mosaic patches induced by allo-DNA (2.94%) was about three times higher than that observed among flies treated with iso-DNA (1.01%). Chi square analysis in 2 X 2 contingency tables disclosed that allo-DNA induced a highly significant increase in the frequency of mosaics both in comparison with iso-DNA (χ^2 = 36.42) and with Ringer solution (χ^2=62.14). The difference between the effects of iso-DNA and Ringer solution was not significant (χ^2=0.49).

The frequency of mosaic patches for individual loci among flies treated with allo-DNA ranged from 0.5% *(st)* to 2.08% *(sn)*. The *white (w)* locus constituted a special case; while no w^+ mosaics were observed in the treated generation, they were observed in subsequent generations. In any event, the loci used in these experiments were not of equal value in the scoring of mosaicism or in estimation of the size of mosaic patches, making comparisons of specific locus rates somewhat uncertain.

As might be expected, flies exhibiting more than one mosaic patch were very rare. Of the 260 allo-DNA treated mosaic flies which form the basis of Table 3, two exhibited two separate mosaic patches. In both cases the patches differed in the locus altered. More pertinent, in experiments where y sn^3 embryos were treated with $y^+ sn^+$ DNA, it was possible to observe alterations for both loci in the same patch. One such case was observed among 3,483 flies. Available data (Fox and Yoon, 1970) yielded a frequency of 0.017 for changes of y to y^+, and a similar frequency for changes of sn^3 to sn^+. Thus, if the two events occured independently, the frequency of their joint occurrence should be 0.000289. Since the observed frequency of the double event was 0.000287, it may be concluded that double events affecting the same cell lineage occurred no more frequently than expected by chance.

The data pertaining to changes from dominant allele to recessive *(ru, h, st, cu, ca)* were deficient in that the design of the experiments from which they came did not allow tests with iso-DNA. Statistical analysis showed a highly significant difference between the effects of allo-DNA and Ringer solution (Fox and Yoon, 1966). Since, however, the specificity of the effects requires comparison with those of iso-DNA, the conclusion that allo-DNA induced changes from dominant to recessive needs to be qualified. This problem merits further study.

The gene *vermilion (v)* is of special interest because it is non-autonomous in its phenotypic effects (Sturtevant, 1920; Beadle and Ephrussi, 1937) and affords a particularly useful amplifying system. Wild-type eye color in *Drosophila* results from the presence of two classes of pigments, the brown ommochromes and red pterins. The mutant v^1 suffers from the inability to convert tryptophan to kynurenine, which is the precursor of the ommochromes (Butenandt, 1953).

TABLE 3

*Frequency of mosaic patches observed for individual autonomous loci.**

Treatment	Locus										
	y^+	sc^+	w^+	cv^+	sn^+	$ru^{\ddagger}$	$h^{\ddagger}$	$st^{\ddagger}$	$cu^{\ddagger}$	$ca^{\ddagger}$	*Total*
Allo-DNA	0.0164 (5061)	0.0014 (5666)	0 (1173)	0.0003 (5666)	0.0208 (4989)	0.0141 (5666)	0.0079 (5666)	0.0005 (5666)	0.0011 (5666)	0.0011 (5666)	0.0294 (9149)
Iso-DNA	0.0064 (4659)		0.0011 (1839)		0.0066 (4731)						0.0101 (3153)
Ringer Solution	0.0030 (3357)	0 (1405)	0 (980)	0 (1405)	0.0045 (3357)	0.0085 (1405)	0.0028 (1405)	0 (1405)	0 (1405)	0 (1405)	0.0086 (4762)

*Data from Fox and Yoon (1966) and Fox and Yoon (1970). Number of flies observed given in parentheses.
+Direction of change: recessive to dominant.
‡Direction of change: dominant to recessive.

The defect results from the absence of tryptophan pyrrolase activity (Baglioni, 1960); the *vermilion* locus includes the structural gene for the enzyme (Tartof, 1969). This conversion is normally carried out in a variety of tissues outside of the eye. As shown by Sturtevant and by Beadle and Ephrussi, a *vermilion* individual possessing a v^+ mosaic patch outside of its eye will frequently develop brown pigment in its v eyes as a result of the diffusion of kynurenine from the v^+ tissue. The system may be sensitized by inclusion of the autonomous mutant *brown (bw)* in the genotype of treated flies. This mutant suffers from the inability to produce pterins. In combination with v^1 it results in a slightly tinged white eye, making possible the detection of very small amounts of brown pigment in v^+ mosaics.

Treatment of v^1 *;bw* embryos with v^+ allo-DNA results in the production of flies with colored eyes (Fox and Yoon, 1970). The eye color exhibited by these affected flies is variable from fly to fly, but it affects the whole eye and is bilaterally symmetrical. Both the shade and intensity of pigmentation vary, and an objective system of color classification has been developed (Fox *et al.*, 1975). At one end of the range, affected flies may be as light-eyed as the most deeply tinged unaltered v^1 *;bw,* while at the other end they may be almost full-browns but never as dark as v^+ *;bw.* The classification system is such that flies are designated *colored* only if they are definitely darker in eye color than those which fall in the range of overlap; all others are classified as white. The interpretation of this variability is that it depends on the size and location of the induced v^+ patch and the time and concentration of kynurenine production relative to the period of pigment deposition in the eye, which remains v^1 in genotype. Occasionally flies are seen with discrete colored sectors in the eye itself; in such cases, the v^+ patch is in the eye rather than elsewhere.

Table 4 contains a summary of experiments in which v^1 *;bw* embryos were treated with allo-DNA, iso-DNA, and sucrose. As in experiments with autonomous loci, allo-DNA yielded altered flies (colored) with a frequency which was significantly higher ($P \backsim 0.002$) than that observed for iso-DNA, in this case almost nine times higher. Iso-DNA yielded a higher frequency of colored flies than did sucrose, but the difference was not significant.

2. *The effect of allo-DNA is locus-specific, and is related solely to its genetic content.*

The specificity of the effects of allo-DNA, as compared with iso-DNA, suggests a relationship to the difference between its genetic content and the genotype of the treated embryos. A more adequate test of this specificity would come from experiments where *a b* embryos were treated with *a B* DNA on the one hand, and *A b* DNA on the other. Each of these DNA's would be half-iso and half-allo, and differential effects could be attributed to real gene specificity.

TABLE 4

*Number and frequency of colored flies resulting from treatment of v^1;bw embryos.**

Treatment	No. flies examined	No. colored (Frequency)
Allo-DNA *(v^+;bw^+)*	1533	12 (0.0078)
Iso-DNA *(v^1;bw)*	3452	3 (0.0009)
Sucrose	3503	1 (0.0003)

*Data from Fox and Yoon, 1970 and Fox *et al.*, 1975.

The results of an experiment in which *y* sn^3 embryos were treated with y^+ sn^3 DNA or *y* sn^+ DNA have been published (Fox and Yoon, 1970). Treatment with y^+ sn^3 DNA produce an excess of y^+ mosaics, while treatment with *y* sn^+ DNA produced an excess of sn^+ mosaics. If the responses of the two loci were independent of the genetic content of the DNA's, results like those observed would be expected to occur by chance with a probability of about 0.001.

3. *The induced alterations are sometimes transmitted to subsequent generations, both by affected and unaffected flies.*

Transmission to subsequent generations must be demonstrated if the alterations induced in treated flies by allo-DNA are to be regarded as genetic. Developmental considerations outlined above predict that not all altered flies will transmit the effect, and that some unaltered flies will produce altered progeny.

Table 5 contains data which illustrate the frequency with which treated flies produce affected progeny. The frequency of affected progeny from allo-DNA treated parents was 3 to 4 times higher than that exhibited by progeny of iso-DNA or Ringer solution treated parents. This difference is highly significant in both cases ($P<0.001$), while the difference between the F_1 of iso-DNA treated and Ringer solution treated flies is not significant. Similar data have been published for the remaining autonomous loci listed in Table 3 (Fox and Yoon, 1966), and for the non-autonomous v^1 ;*bw* system (Fox and Yoon, 1970).

Several additional features of the results summarized in Table 5 are important. The first is that both mosaic and unaltered parents were tested. None of the mosaic iso-DNA or Ringer solution treated parents produced altered progeny. On the other hand, 530 allo-DNA treated non-mosaics were tested, producing 53 (0.0016) mosaics among 32,965 F_1 progeny; 36 allo-DNA treated mosaics produced 19 (0.0082) mosaics among 2,318 progeny. Thus, mosaics occurring among iso-DNA or Ringer solution treated flies did not transmit their alterations, presumably because they are strictly somatic, while both mosaic and

TABLE 5

*Affected progeny of treated y w sn^3 parents.**

Treatment	Parents No.	Mosaics in F_1 y^+	w^+	sn^+	Total	No. of F_1 examined
Allo-DNA (y^+ w^+ sn^+)	566	37 (0.0010)	20 (0.0006)	15 (0.0004)	72 (0.0020)	35,283
Iso-DNA (y w sn^3)	450	5 (0.0002)	3 (0.0001)	8 (0.0003)	16 (0.0007)	23,986
Ringer	297	4 (0.0002)	3 (0.0001)	3 (0.0001)	10 (0.0005)	20,455

*Data from Fox and Yoon, 1966. Frequencies given in parentheses.

non-mosaic allo-DNA treated flies exhibited germ-line alterations. The *white* locus constitutes a special example of this last point. Although no w^+ mosaics were observed among allo-DNA treated flies themselves (Table 3), 20 were observed among their F_1 progeny (Table 5).

The above data seem to exhibit a weak correlation between mosaicism in the allo-DNA treated generation and that observed in the F_1. The frequency of mosaics in the F_1 produced by mosaic parents (0.0082) was significantly higher than that among F_1 produced by non-mosaic parents (0.0016). Furthermore, 24 y^+ mosaics yielded 13 y^+ and 5 w^+ mosaic progeny (out of 1,614), while 12 sn^+ mosaics yielded 1 sn^+ mosaic and no y^+ or w^+ mosaics (out of 704). Thus, there seems to be a tendency for allo-DNA treated mosaics to transmit their own type of mosaicism.

Finally, an explicit statement of the most striking observation regarding transmission of the effects of allo-DNA needs to be made. All affected F_1 flies were mosaics; no whole-body alterations were observed.

4. *When transmission does occur, a degree of uncertainity is frequently observed during the first few generations. This uncertainty consists of instances where the introduced information appears to be expressed but not transmitted, and of instances where it is transmitted but not expressed. The result is apparently non-Mendelian patterns of transmission during this interval.* (Fox and Yoon, 1966; Fox and Yoon, 1970).

Some, but not all, of affected F_1 flies transmitted the induced alterations to subsequent generations. Table 6 illustrates this observation. Thus, of the 37 y^+ F_1 mosaics listed in Table 5, 26 were tested by matings with untreated *y w sn*^{3} flies; 17 F_2 y^+ mosaics and 2 F_3 y^+ mosaics were tested in a similiar fashion.

TABLE 6

*Transmission by F_1 y^+ mosaics of allo-DNA induced effects.**

Mosaics tested		Mosaics among progeny			No. progeny
Generation	Number	Generation	y^+	w^+	examined
F_1	26	F_2	24	1	1,214
F_2	17	F_3	10	1	544
F_3	2	F_4	1	0	150

*Data from Fox and Yoon, 1966.

It is apparent that these data do not illustrate Mendelian transmission. They do however, illustrate two features which appear to be generally characteristic of the early transmission of allo-DNA induced effects. The first is that affected flies appearing in subsequent generations were observed. The second is that w^+ mosaism appeared in the F_2 and F_3 of lines in which it had not previously been evident; a delay of two or three generations was manifested between the time of treatment with allo-DNA and the first appearance of its effects. Similar results have been observed for the transmission of the effects of allo-DNA on the other autonomous loci listed in Table 3 (Fox and Yoon, 1966).

The difficulties of detecting mosaicism render experiments with autonomous genes extremely laborious, and more adequate genetic analysis is possible with the v^1 *;bw* system (Fox and Yoon, 1970; Fox *et al.*, 1971a). Table 7 summarizes results obtained in the early generations of a lineage which ultimately gave rise to a number of "v^+ transformed" stocks exhibiting sex-linked recessive alterations.

The first enigma presented by this lineage is that colored flies do not appear until the F_4 generation. Since the genetic alteration responsible for the colored phenotype was subsequently demonstrated to be sex-linked and recessive, some of the white-eyed females observed during the first three filial generations could have been heterozygous, and should have yielded colored sons; none of these were observed. Furthermore, white-eyed, F_3 males produced colored daughters in the F_4 generation. The latter would be presumed to be homozygous for the induced alteration, in which case they should have received an X chromosome carrying the alteration from their father. The hemizygous fathers, therefore, should have been colored. Thus, the induced alteration appears to have been transmitted through four generations without being expressed.

A further enigma is presented by the transmissional behavior of F_4 colored flies mated with each other. Since the transmission of the alteration is sex-linked

TABLE 7

*Early transmission of the effects of $v^+;bw^+$ DNA by treated $v^1;bw$ flies.**

Mating	Generation Produced	Progeny White-eyed ♂♂	Progeny White-eyed ♀♀	Progeny Colored ♂♂	Progeny Colored ♀♀
Mass: P white-eyed ♂♂⁺ X P white-eyed ♀♀⁺	F_1	1635	1370	0	0
Mass: F_1 white-eyed ♂♂ X F_1 white-eyed ♀♀	F_2	398		0	0
Single pair: F_2 white-eyed ♂ X F_2 white eyed ♀	F_3	40	34	0	0
4 pairs: F_3 white-eyed ♂♂ X F_3 white-eyed ♀♀	F_4	63	101	31	11

*Data from Fox and Yoon, 1970.
+Treated with allo-DNA $(v^+;bw^+)$. P generation consisted of 85♂♂ and 93♀♀, all white-eyed.

recessive in the transformed stocks from this lineage, such matings should produce nothing but colored progeny. All of the seven single-pair matings performed with F_4 colored flies yielded white-eyed progeny (Table 8). Two of the seven matings yielded results suggesting that the female parents were heterozygous, in spite of their colored phenotype. Three matings yielded results suggesting that the colored male parents did not transmit the alteration. The remaining two matings yielded results suggesting that neither parent transmitted the alteration.

It should be noted that not all lineages exhibit this initial instability. Some lineages exhibit Mendelian transmission from the very beginning, sometimes with sporadic irregularities during the first few generations and sometimes with no irregularities (Fox and Yoon, 1970).

5. *Transmission may be stabilized after the initial period and transformed stocks may be established. Thereafter, strict Mendelian transmission is observed* (Fox and Yoon, 1970; Fox *et al.*, 1971a).

In many lineages, if not most, the instabilities exhibited during the early generations result in loss of the allo-DNA induced effect. In some lineages, however, stabilization occurs and altered stocks may be established. Thus, in the example given in Tables 7 and 8, stabilization occurred after the F_4 generation. Progeny tests of the F_5 flies showed that they transmitted the alteration in a Mendelian fashion in all cases: white-eyed males were hemizygous for an unaltered X, colored males were hemizygous for an altered X, white-eyed females were heterozygous with one altered and one unaltered X, and colored females were homozygous with two altered X chromosomes. It was therefore possible to establish homozygous, altered ("transformed") stocks. Similar results have been

TABLE 8
*Aberrant transmission by F_4 colored flies.**

Class	No. of single-pair matings	F_5 progeny White-eyed ♂♂	♀♀	Colored ♂♂	♀♀
1	2	38	27	33	39
2	3	75	143	46	0
3	2	84	118	0	0

*Data from Fox and Yoon, 1970.

achieved with other lineages (Fox and Yoon, 1970), and extensive analysis has shown that these "v^+ transformed" stocks behave in a Mendelian fashion (Fox *et al.*, 1971a).

6. *Conventional genetic mapping techniques demonstrate that the induced alterations present in established stocks are located at defined chromosomal sites. The genetic alteration may map either at the homologous chromosomal site or at a heterologous site.*

Table 9 gives the map position of the genetic alterations responsible for the colored phenotype exhibited by thirteen "v^+ transformed" stocks. Those stocks designated as "primary" were derived as described above from v^1 ;*bw* flies which were treated with v^+ allo-DNA. The remaining stocks were derived from primary stocks in a variety of ways described in the cited references.

The striking feature of these mapping results is that the genetic alterations in "v^+ transformed" stocks map at two distinct locations. One group of stocks is characterized by an alteration which maps at 33.0 on the X (lst) chromosome; a second group exhibits an alteration mapping at 0.0 on the X chromosome; one stock exhibits two separable alterations, one at 0.0 and the second at 33.0. The position 1-33.0 is that of the *vermilion (v)* locus itself (Lindsley and Grell, 1968), which includes the structural gene for tryptophan pyrrolase (Tartof, 1969). The position 1-0.0 is in a region which includes the *suppressor-of-sable [su(s)]* locus (Lindsley and Grell, 1968). Mutants of *su(s)* suppress some *v* mutants, including v^1 , by restoring partial tryptophan pyrrolase activity (Baglioni, 1960).

No conclusion is yet possible about the relative frequencies with which alterations are induced at these two locations in v^1 ;*bw* flies treated with v^+ DNA. Although *6203♂4* and *8540♂10* stocks are of independent origin, the four primary stocks with alterations located at 1-0.0 may represent as many as four or as few as one primary event.

As will be shown in succeeding sections, all of these stocks exhibited properties suggesting that their altered phenotype resulted from the introduction of

TABLE 9
Map positions of genetic alterations in "v^+ transformed" stocks.

Map position	Stock number	Source	Reference
	6203♂4	Primary	Fox and Yoon, 1970
	8540♂10	Primary	Fox and Yoon, 1970
1-33.0	*col-2*	Isolated from *e♀6dom* by crossing-over	See text
	col-3	From *e5·6* by apparent transposition	unpublished data
	e5·6	Primary	Fox and Yoon, 1970
	e5·4ʳ	Primary	unpublished data
	e♀6	Primary	unpublished data
	e14	Primary	unpublished data
1-0.0	*e14ʳ*	By selection from *e14*	unpublished data
	e14ʸ	By selection from *e14*	unpublished data
	F_3♂2	Acridine orange treatment of *e14ʳ*	unpublished data
	col-1	Isolated from *e♀6dom* by crossing-over	See text
1-0.0 and 1-33.0	*e♀6dom*	Acridine orange treatment of *e♀6*	See text

v^+ information at the time of treatment with allo-DNA. In-so-far as this conclusion could be justified, the mapping experiments showed that the introduced information may be located either at the homologous chromosomal site or at a heterologous site, or at both.

7. *The molecular mechanisms involved in the expression of alterations located at heterologous sites appear to differ from those located at homologous sites.*

This statement is not yet fully documented, but is supported by evidence which is at least strongly suggestive.

The colored phenotype exhibited by all "v^+ transformed" stocks, regardless of the map position of the induced alteration, resulted from the presence of variable, but less than wild-type, amounts of ommochrome demonstrable by chromatography of extracted pigments (N. Slepekis, unpublished). Their larval fat bodies exhibited kynurenine-flourescence (T. M. Rizki, personal communication). Extracts of adults exhibit tryptophan pyrrolase activity ranging from about 2% to 37% of wild-type (Tobler, 1975). Insertion of mutants like *scarlet (st)* into their genome blocks ommochrome production (Fox, unpublished); *st* blocks the chain of reactions leading to ommochrome subsequent to the step catalyzed by tryptophan pyrrolase (Beadle and Ephrussi, 1937). All "v^+ transformed" stocks therefore exhibit partial restoration of v^+ gene activity.

The interpretation, suggested for the above observation noting that the restoration of v^+ activity is partial rather than complete is that flies in these stocks are phenotypic (but not genetic) mosaics, with v^+ activity expressed only in mosaic patches. This interpretation is supported by the resemblance of their phenotype (Fox, unpublished) to that produced by $T(1;2)v^{65b}$ a translocation resulting in a V-type position effect of *vermilion,* that is, phenotypic mosaicism (Lefevre, 1969).

There are important differences, however, between those "v^+ transformants" with alterations localized at 1-33.0 and those localized at 1-0.0. In the first place, the former are uniformly *dominant,* in the sense that females heterozygous for the altered X chromosome and an X carrying v^l, but homozygous for *bw,* exhibit colored eyes. The latter, on the other hand, are uniformly *recessive;* females heterozygous for the altered X and an X carrying v^l, but homozygous for *bw,* exhibit off-white eyes like those of unaltered v^l ;*bw.*

The meaning of this observation may be better understood, perhaps, if the complete genotypic designation of these heterozygotes is given. Let the symbol v^* designate the genetic alteration at 1-33.0 in "dominant transformants", and let *col* be the symbol for the alteration at 1-0.0 in "recessive transformants". Then heterozygotes for "dominant transformants" would have the genotype

$$\frac{su(s)^+ \; v^*}{su(s)^+ \; v^l} ; \frac{bw}{bw} ,$$

while heterozygotes for "recessive transformants" would be

$$\frac{col \quad v^l}{su(s)^+ \; v^l} ; \frac{bw}{bw}$$

In the former case, the term *dominant* refers to the relationship between v^* and v^l ; in the latter case, the term *recessive* refers to the relationship between *col* and $su(s)^+$.

We have a whole series of observations indicating a close functional relationship of the genetic alterations in "recessive transformants" to the *su(s)* locus and shall publish these more fully (Fox, Kreber, Liu, and Yoon, in preparation), together with a set of observations pertaining to the functional relationship of "dominant transformants" to the *v* locus, in a separate paper (Yoon, Kreber, and Fox, in preparation). For our present purposes it is sufficient to suggest that allo-DNA induced alterations localized at 1-33.0 restore tryptophan pyrrolase activity by means of molecular mechanisms resembling those associated with v^+ (which is dominant), while alterations localized at 1-0.0 restore tryptophan pyrrolase activity by mechanisms resembling those of *su(s)* mutants (which are partially or completely recessive). Further interpretation of this suggestion will be given below.

8. *The altered phenotypic effect is lost in established "transformed" stocks with a frequency in gametes approximating 10^{-4}. When such loss occurs, the phenotypic effects then observed suggest that the allele originally present at the altered site is still there* (Fox *et al.*, 1971a).

Although no flies in "v^+ transformed" stocks exhibited the full brown-eyed phenotype characteristic of authentic $v^+;bw$, many such stocks exhibited phenotypic overlaps with $v^l\ ;bw$. Such white-eyed flies almost always produced the same array of colored progeny as did their colored siblings, indicating that their lack of ommochrome was not attributable to loss of the genetic alteration originally induced by v^+ DNA. Actual loss, however, was signaled by the occurrence of genetically white-eyed flies in these stocks. When tested, these flies produced no colored progeny. Their genetic behavior, in all respects, was identical to that of the untreated $v^l\ ;bw$ stock.

Genetically, such losses occurring in "dominant transformants" may be symbolized as follows:

$$su(s)^+\ v^*\ ;bw \longrightarrow su(s)^+\ v^l : bw$$

In "recessive transformants", they would be symbolized:

$$col\ v^l\ ;bw \longrightarrow su(s)^+\ v^l\ ;bw$$

Their frequency (10^{-3} to 10^{-4}) was, however, several orders of magnitude higher than that of spontaneous mutations of v^+ to v or *su(s)* to $su(s)^+$, and mutationally unstable alleles are not known at these two loci (Lindsley and Grell, 1968). Furthermore, there was no evidence that the loss of pigment-producing capacity in these white-eyed flies was accompanied by low viability, abberant phenotypes, or any chromosomal abnormality.

Unfortunately, it has not yet been possible to test for positively identifiable products (RNA or protein) of either v^l or $su(s)^+$ in these white-eyed revertants. Thus rigorous proof is lacking, but the negative evidence strongly suggests that the v^l gene in "dominant transformants", and the $su(s)^+$ gene in "recessive transformants", have remained unchanged throughout the entire process of alteration to a colored phenotype and reversion to white.

9. *In spite of a large-scale search in specially designed experiments, whole-body "transformants" have never been observed among the progeny of allo-DNA treated flies. Even in "transformed" stocks, whole-body "transformants" are never observed. Although flies from these stocks transmit the induced alteration*

through all of their gametes (except when loss occurs), they are always phenotypic mosaics. The phenotypic effects of the induced alteration are expressed in some tissues, and the effects of the original chromosomal gene are expressed in other tissues.

That affected progeny of allo-DNA treated flies are invariably phenotypic mosaics has already been noted. Table 10 summarizes the results of several experiments in which a large-scale search was made for whole-body transformants (Fox *et al.*, 1970). Some of these experiments included the use of genetic or biochemical selective systems which would make possible the detection of whole-body transformants without detailed visual inspection of each fly. Others included detailed visual inspection. In all, approximately 361,245 gametes from allo-DNA-treated flies were tested and not one gave rise to a verifiable whole-body transformant.

TABLE 10

*Systems used in the search for whole-body transformants**

Allo-DNA-induced changes	Methods of detection of whole-body changes	Number of gametes tested
Lethal reversion: $l \rightarrow l$	Balanced lethal system *allowing survival only of* l^+ revertants	68,233†
Rosy reversion: $ry^2 \rightarrow ry^+$	Purine killing of all but ry^+ revertants	116,050†
Changes to wild-type:	Visual inspection	
$y^1 \rightarrow y^+$		38,283
$sn^3 \rightarrow sn^+$		25,284
$bw \rightarrow bw^+$		41,841
$st \rightarrow st+$		34,838
$v \rightarrow v^+$		15,501
$cn \rightarrow cn^+$		8,600
$ry^2 \rightarrow ry^+$		9,623
$mal^{bz} \rightarrow mal^+$		1,501
$y^2 \rightarrow y^+$		1,501

*Data from Fox *et al.*, 1970.
†Estimated.

These data must be qualified by noting that some of the target genes were previously listed among those which have failed to respond to treatment with allo-DNA. This is to say that no changes were observed in the treated generation itself, but it does not exclude the introduction of pertinent DNA fragments by the treatment. On the other side of the coin, we must include perhaps 10^7 flies observed thus far in the "v^+ transformed" stocks, none of which can be regarded as *bona-fide* whole-body "transformants". None have exhibited the true brown eye-color expected in a whole-body $v^+; bw$ individual.

Another experiment directed to this point was one in which an attempt was made to achieve integration of pieces of allo-DNA into recipient chromosomes by enhancing the frequency of exchange through the application of X-irradiation (Fox *et al.*, 1970). Eggs of the genotype $w^{a2}\, sn^3$ were treated with w^+sn^+DNA, allowed to hatch, and exposed to 1200 r of X-rays at the end of the first larval instar. Becker (1957) and Baker (1967) have shown that at this stage of development the eye primordium contains some 16 to 20 cells which give rise to defined regions of the adult eye. The objective of our experiment was to see if X-irradiation induces exchanges between introduced DNA and the chromosome, producing the expected patterns of the eye mosaicism for $w^{a2} \rightarrow w^+$ changes. In addition, since the germ line was also irradiated, the next generation was examined for whole-body $w^{a2} \rightarrow w^+$ or $sn^3 \rightarrow sn^+$ changes.

A total of 1,246 treated flies, representing a sample of more than 20,000 primordial eye cells, were examined for the type of mosaicism which would result from induced exchange. None were found. In addition, 10,342 progeny of treated flies were examined for whole-body changes. None were found. Thus, even under conditions favorable for exchange, no evidence for the occurrence of whole-body "transformants" was encountered.

All observed "transformants", therefore, were phenotypic mosaics in which the effects of the induced alteration were expressed in some tissues and the effects of the original chromosomal gene were expressed in other tissues. This was additional, strong evidence for the persistence of the original allele at the mapped position of the alteration, but rigorous proof will still demand demonstration of immediate RNA or protein product.

10. *Coded studies have disclosed the presence of cytological alterations in the salivary chromosomes of "transformed" stocks. In every stock examined, except untreated controls, extra chromatin has been observed in a small proportion of salivary gland chromosomes, associated with the chromosome band corresponding to the map position of the genetic alteration.*

Careful studies have been performed on the salivary chromosomes of "v^+ transformed" stocks with alterations mapping either at 1-0.0 or 1-33.0 (Fox and Valencia, 1975). In initial studies, the nontransformed control stock and six "transformed" stocks were assigned code numbers prior to cytological examination. Salivary chromosome regions 1B9 • 10 - 1C2 • 3 and 10A1 • 2 - 10B1 • 2 were carefully examined in each of the coded stocks. (Band 1B11 was tentatively identified as the site of the *suppressor-of-sable* locus at 1-0.0 and band 10A1 was the site of the *vermilion* locus at 1-33.0). When cytological studies were complete, each of the stocks was identified. In every "transformed" stock examined (Table 11), anomalies had been scored in association with the chromosome region corresponding to the map position of the DNA-induced alteration. In the control stock, anomalies were observed at neither position.

TABLE 11

*Genetic and cytological positions of DNA-induced alterations in control and "transformed" stocks**

Stock	Map position of DNA-induced alteration	Region of cytological anomalies
v^1 ;*bw*	None	None
6203♂4	1-33.0	10A1•2
8540♂10	1-33.0	10A1•2
e♀6	1-0.0	1B11
e5•6	1-0.0	1B11
e♀6dom	1-0.0 and 1-33.0	1B11 and 10A1•2
col-2	1-33.0	10A1•2

*From Fox and Valencia, 1975.

Approximately 10-15% of the nuclei in transformed stocks exhibited significant departure from normality in the pertinent chromosome region. The perturbations ranged from minor alterations of banding pattern to apparent pieces of extra chromatin in the most extreme cases. In stocks with "v^+ transformations" mapping at 1-33.0, apparent extra chromatin was observed with frequencies varying from 1/200 to 1/500. In these cases the abnormal structures were associated with salivary band 10A1 • 2, forming a band-like structure, frequently an almost perfect doublet (open or closed), often lying in an abnormal position, and with fine chromatin threads connected to the chromosomal doublet. In stocks with "v^+ transformations" mapping at 1-0.0, normal structures were frequently thread-like, lying to the side or off the tip of the chromosome, with compact regions which sometimes resembled chromomeres, and with fine threads connected to the chromosomal 1B11 region.

11. *DNA prepared from "transformed" stocks induces "transformations" resembling those induced by the original allo-DNA.*

In bacterial transformation systems, DNA prepared from transformed cells is capable of inducing transformations like the original. A similar test of the activity of DNA from "v^+ transformed" stocks has been performed (Fox *et al.*, 1975).

In these experiments, v^1 ;*bw* embryos were subjected to five different treatments: sucrose, iso-DNA *(*v^1 *;bw),* allo-DNA *(*v^+*;bw*$^+$*),* and the DNA's prepared from two different "v^+ transformed" stocks *(6203♂4,* which localizes at 1-33.0, and *e5•6,* which localizes at 1-0.0). The results are given in Table 12.

TABLE 12

*Number and frequency of colored flies resulting from treatment of v^1 ;bw embryos with iso-DNA, allo-DNA, and "transformed" DNA's**

Treatment	No. flies examined	No. colored	Frequency colored
Sucrose	3,095	1	0.0003
Iso-DNA *(v^1;bw)*	2,378	3	0.0013
Allo-DNA *(v^+;bw^+)*	1,132	9	0.0079
6203♂4	3,217	156	0.0485
e5•6 DNA	2,926	57	0.0195

*Data from Fox *et al.*, 1975.

As in previous experiments, treatment with iso-DNA yielded a higher frequency of colored flies than did sucrose, but as in previous work, the difference was not significant (P=0.32). Allo-DNA yielded colored flies with a frequency similar to that observed in previous work, some 6-7 times higher than that of iso-DNA. This difference, which is highly significant (P=0.003), measured the "transforming" activity of allo-DNA.

6203♂4 DNA produced colored flies with a frequency 38-39 times greater than that for iso-DNA and 6 times higher than that for allo-DNA. The other transformed DNA, *e5 • 6*, produced colored flies with a frequency about 5 times higher than that for iso-DNA and 2.4 times higher than for allo-DNA. In both cases the yield of colored flies was significantly higher than that of allo-DNA ($P=8.2 \times 10^{-10}$ and $P=8.1 \times 10^{-3}$, respectively), while the yield obtained with *6203♂ 4* DNA was significantly higher than that obtained with *e5 • 6* DNA ($P=5.4 \times 10^{-10}$).

Of the 156 colored flies resulting from treatment with *6203♂4* DNA, 142 were observed in a single experiment. Omitting this experiment, the total yield of *6203♂4* DNA was 2,535 white-eyed and 14 colored flies. This result differed significantly from that obtained with v^1 *;bw* DNA (P=0.01), but did not differ from that obtained with allo-DNA (P=0.37).

Similarly, of the 57 colored flies resulting from treatment with *e5• 6* DNA, 42 were observed in a single experiment. Omitting this experiment, the total yield of *e5 • 6* DNA was 2,808 white-eyed and 15 colored flies. This result differed significantly from that obtained with v^1 *;bw* DNA (P=0.02), but not from that obtained with allo-DNA (P=0.37). Omitting the two exceptional experiments, the effects of *6203♂4* and *e5• 6* DNA were not significantly different from each other.

The clear result of this work is that DNA from "transformed" stocks was as effective as allo-DNA in the production of colored flies, or more effective if the cases of extreme clustering are included. Thus these results suggest that v^+ information, originally introduced by treatment with allo-DNA, was present in flies of "transformed" stocks many generations after their origin. It follows that colored flies resulting from treatment with "transformed" DNA should sometimes transmit the effect to subsequent generations, but this has not yet been tested, nor has the localization of the alterations induced by "transformed" DNA been determined.

The occurrence of a few large bursts of "transformants" in isolated experiments with transformed DNA is not presently understood. There are many possibilities for variability in the efficiency of "transformation" inherent in the present experimental system, and future work needs to be directed toward the establishment of conditions under which "transformation" is uniformly high. Possibilities for the accomplishment of this objective are discussed below.

12. *Some, if not most, of the apparent losses of the allo-DNA induced alterations from the mapped sites in "transformed" stocks, described above, apparently involve transposition of the introduced alteration to a new site in the genome. In each case the original evidence for a change in position is genetic, but in all cases examined cytological confirmation has been obtained of actual physical transfer of the cytological anomaly associated with the original site.*

The loss of the altered phenotypic effect in established "v^+ transformed" stocks (point 8, above) constitutes a kind of instability or metastability in an otherwise genetically stable situation. The relatively high frequency of these events (approximately 10^{-4}) could, in a formal sense, characterize the genetic condition induced by allo-DNA as highly mutable. Since intensive study of "mutable alleles" in maize has disclosed that their mutability is attributable to their association with transposable elements which change their position in the genome with high frequency (Fincham and Sastry, 1974), we were led to consider the possibility that the genetic alterations induced by allo-DNA could undergo transposition. Evidence has been accumulated for three kinds of apparent transpositions.

a) *Transposition from a heterologous site (1-0.0) to the homologous chromosome position (1-33.0).* - Two apparent instances of this type of transposition have been encountered (Yoon, Kreber, and Fox, unpublished). One occurred among the male progeny of y^1 *e5•6* gt^{E6} *cv* v^1 *f* / v^1 ; *bw* / *bw* females (for identity of gene symbols see Lindsley and Grell, 1968; *e5•6* is the primary "recessive transformant" listed in Table 9). One male was encountered which was colored and forked *(f)* in phenotype, but which exhibited an eye color more closely resembling that of "dominant transformants" than of *e5•6*. A homozygous stock possessing this altered X chromosome was established and subjected to genetic analysis, the results of which demonstrated that the genetic alteration responsible for the colored phenotype was now dominant and mapped at 1-33.0. In its new position, the alteration was designated as *col-3* (Table 9).

A minimum of two events are required to account for these changes: a crossover between the two X's in the heterozygous female, separating *f* from the other markers, and a transposition of the *e5 • 6* genetic element from 1-0.0 to 1-33.0. Cytological and genetic analysis is in progress to test this interpretation.

The second instance of such an apparent transposition occurred in an experiment in which flies of the *e♀6* stock were grown on medium containing acridine orange (5ug/ml). In the seventh generation of successive transfers on this medium an individual male was produced with eyes more deeply colored than usual observed in the *e♀6* stock: its phenotype resembled that exhibited by the dominant *6203♂4* stock. The X chromosome of this male was rendered homozygous, and out-crosses to v^1 *;bw* demonstrated that its phenotypic effect was dominant, unlike the X of *e♀6* but like that of *6203♂4* and *8540♂10*. The stock homozygous for this X is designated *e♀6dom* (Table 9).

A mapping experiment was performed by examining the progeny of females heterozygous for the *e♀6dom* X chromosome and an X marked with the mutants y^2 *cv* v^1 *f*. This experiment disclosed that the *e♀6dom* chromosome carried two genetic alterations, one at 1-0.0 and the other at 1-33.0. The alteration at 0.0 was designated *col-1*, and that at 33.0 was designated *col-2* (Table 9). The two elements were separated by crossing-over and tested individually. The alteration *col-2* was dominant and exhibited a cytological anomaly at 10A1 • 2 in salivary chromosomes (Table 11).

The *e♀6dom* chromosome may have arisen as a result of an extra duplication of the *e♀6* element and transposition of the extra copy from 1-0.0 to 1-33.0. Similar cases have been observed for transposable elements in maize (Von Schaik and Brink, 1959). The role of acridine orange, if any, is not understood.

b) *Transposition from the homologous chromosome position (1-33.0) to some heterologous site (unidentified), followed by retransposition to the homologous location.* Systematic studies have been performed of events affecting *6203♂4*, *8540♂10*, and *col-2*, all of which are located at 1-33.0 and exhibit cytological anomalies associated with salivary chromosome bands 10A1 • 2 (Fox, Kreber, and Valencia, unpublished). Three types of events have been encountered: apparent loss, apparent transposition, and apparent inactivation and reactivation *in situ*. One case will be described as an illustration.

In cases involving *col-2*, the following cross was performed:

$$\frac{sn^3 \quad lS12 \quad + \quad col\text{-}2 \quad df(1)m^{259-4}}{+ \quad + \quad lQ54 \quad df(1)v^{166} \quad m} \; X \; \frac{lQ54 \quad df(1)v^{166} \quad m}{v \cdot Yy^+}$$

(for identity of gene symbols see Lindsley and Grell, 1969; and Lefevre, 1971).

All daughters of this cross should have the same genotype as their mothers (barring easily detected crossovers and non-disjunctionals), and should be colored in phenotype unless an event occurred affecting the *col-2* element. Among 25,000 such daughters, one was encountered which was white-eyed. Tests demonstrated that the affected chromosome still carried sn^3, *lS12,* and $df(1)m^{259\text{-}4}$. Five derivatives of this chromosome were subjected to intensive genetic and cytological analysis.

One of the five failed to exhibit the chromosomal anomaly originally associated with 10A1 • 2, and has thus far yielded no phenotypic revertants to colored in extensive tests. This may represent a *bona fide* loss of the *col-2* element. Three of the five lacked the cytological anomaly at 10A1 • 2, but repeatedly yielded phenotypic revertants to colored even after 35 generations. In every case the reversion mapped at 1-33.0 and examined revertant X's exhibited cytological anomalies at 10A1•2. In these cases it appeared that the element had been transposed to an unidentified position, where it produced no phenotypic effect, and that it may be retransposed to 10A1•2 with a restoration of its effect on eye color. The remaining derivative retained the cytological anomaly at 10A1 • 2 even though it was phenotypically white-eyed. It yielded repeated reversions to colored, all of which mapped at 1-33.0 and when examined, exhibited the anomaly at 10A1 • 2. In this case it appeared that the element had suffered reversible inactivation. The element in the original white-eyed female appeared to have suffered a destabilizing event which then gave rise to apparent losses, transpositions, and inactivations. Phenomena similar to these have been exhibited by transposable elements in maize (Fincham and Sastry, 1974).

c. *In heterozygotes possessing an alteration at 1-0.0 on one homolog but not the other, possible transfer without crossing-over from the former to the latter.* - Extensive studies of recombination in the vicinity of the genetic alterations characteristic of "recessive transformants" have yielded unconventional results (Fox, Kreber, Liu, and Yoon, in preparation). These studies have utilized all of the recessives listed in Table 9 except *col-1* designated collectively as *col.* The recombination data position *col* as follows:

$$\underline{y \ 0.02 \ col \ 0.23 \ gt \ .}$$

In a region of this length, interference is expected to be complete; double crossovers should not occur, only single crossovers in either of the two intergenic regions *(y-col* or *col-gt).* Analysis of other cases has shown that apparent double crossovers in regions as short as this arise by the non-reciprocal mechanism referred to as gene conversion, rather than by classical reciprocal crossing-over (Stadler, 1973).

Recombination in the *y-col-gt* segment of the X chromosome has been studied among progeny of $col\ /\ y^1\ gt^{E6}$ and $y^1\ col\ gt^{E6}\ /\ +$ females. The data from both crosses exhibit the following features: 1) recovery of reciprocal recombinants between y^1 and *col;* 2) recovery of reciprocal recombinants between *col* and gt^{E6} ; and 3) high negative interference in the *y-col-gt* segment, that is, about 90 times as many double recombinants as expected by chance. These data differ from those obtained in similar crosses involving *bona fide su(s)* mutants instead of *col* elements (Fox and Kreber, unpublished data), where almost all recombination between y^1 and *su(s)* was non-reciprocal, apparently resulting from gene conversion at the *su(s)* site.

The data with *col* elements allow three possible interpretations: 1) recombination in the *y-col-gt* segment results from conventional crossing-over, with high coincidence of crossovers in the two subsegments: 2) recombination results from symmetrical gene conversion at the *col* site (*col* to col^+, and col^+ to *col*), which may be accompanied by single crossovers in either of the adjacent regions: 3) *col* behaves like a transposable element, formally symbolized $su(s)^+ \cdot col$, and recombination in $su(s)^+ \cdot col \,/\, su(s)^+$ heterozygotes results from transposition of the *col* element from homolog to homolog. The first two hypotheses are not easily reconciled with the results obtained with *bona fide su(s)* mutants, but a critical demonstration of the transposition hypothesis remains to be performed.

WORK FROM OTHER LABORATORIES

Two laboratories have initiated independent work with our best-studied system in *Drosophila,* namely our system for the detection of the transfer of the $v+$ gene to $v^1;bw$ embryos. Germeraad (1975) at Irvine and Liembourg-Bouchon (personal communication) in the laboratory of M. Zalokar have injected $v^1;bw$ eggs with DNA prepared from $v+$ flies, and in both cases have succeeded in establishing "$v+$ transformed" stocks resembling ours in all respects. Strikingly, genetic analysis of these stocks has disclosed that the induced alterations were localized at the same two positions as those reported for our stocks; namely at the *vermilion* locus (1-33.0) or at the site of *su(s)* (1-0.0). Work by C. P. Liu in our laboratory has yielded preliminary results confirming the effectiveness of the injection method.

Independent work with *Ephestia* (Nawa and Yamada, 1968), *Neurospora* (Mishra and Tatum, 1973), and *Petunia* (Hess, 1972) tends to confirm some of the effects of allo-DNA listed above. In particular, the points that have received confirmation are 1, 2, 3, 4, 5, and 8. In none of these cases has genetic analysis proceeded far enough to yield evidence bearing on the remaining points, but the exosome model (see below) has been used to interpret the results obtained with *Neurospora* and *Petunia.*

Several reports have been published of apparently successful transfers of the gene for hypoxanthine-guanine phosphoribosyltransferase (HGPRT) into mammalian cells, either by cell fusion (Schwartz *et al.*, 1971) or by the use of isolated metaphase chromosomes (McBride and Ozer, 1973; Burch and McBride, 1975; Willecke and Ruddle, 1975). Metastability was encountered in each of these cases; gene transfer clones tended to lose the donor HGPRT marker in nonselective medium. Such results resemble those described above for *Drosophila.*

INTERPRETATIONS

The complexity of the results reviewed above presents a challenging analytical task. There are three levels at which scientific skepticism is appropriate: That of the experimental results themselves, that of the validity and generality of the conclusions derived from the results, and that of their interpretation. Thus far, avoided interpretation (or labeled where it has been unavoidable), and weaknesses in results or conclusions have been identified. Time will provide the test of their validity.

There are two major problems of interpretation: 1) Does the evidence reviewed here justify the conclusion that gene transfer has been accomplished? and 2) Can a theoretical framework (or model) be constructed which accounts for the complexity of the results and which provides heuristic guidelines? An affirmative answer to the first problem demands critical assurance; theories or models, on the other hand, are subject to change when they no longer account for established experimental results.

By gene transfer we mean that a segment of exogenous DNA, coding for specific genetic information not present in the recipient genome, is established in that genome and perpetuated by replication. In the absence of techniques to directly demonstrate the presence of the introduced nucleotide sequence in the recipient genome, reliance must be placed on detection of the gene's phenotypic effects and on demonstration of its genetic transmission. Unfortunately, little is known about mechanisms controlling gene expression in eucaryotes, and it is usually impossible to know whether a given phenotypic effect is the result of the actual transfer of genetic information or from induced changes in the recipient genome itself. This question can usually be answered only if the immediate products of transcription or translation (RNA or protein) of the introduced nucleotide sequence can be identified. In this most critical sense it cannot be said that gene transfer has been demonstrated in the present work. We had hoped that the v^+ gene and its protein product, tryptophan pyrrolase, might be used for this purpose, but the enzyme has been intractable; methods have not been developed yet to distinguish between tryptophan pyrrolase activity attributable to the product of v+ itself and/or the activated product of a mutant allele. I have avoided interpretation (or labeled it where it has been unavoidable), and weaknesses in results or conclusions have been identified. Time will provide the test of their validity.

Short of this most critical sense, however, the evidence reviewed here constitutes a very strong case for the accomplishment of gene transfer. The specificity and ubiquity of the effects of allo-DNA, the fact that the observed phenotypes are those expected from its genetic content, the transmission of the induced alterations, the establishment of "transformed" stocks, the mapping of the induced alterations by conventional genetic techniques, the demonstration that some of these alterations map precisely at the locus of the target gene, and the "transforming" activity of DNA from "transformed" stocks are observations that are scarcely accountable if they are not the result of gene transfer.

In the case of the "v^+ transformants" there is very strong, although not critical, evidence that the introduced information is that of the v^+ gene itself. In "dominant transformants", the map position of the induced alteration, the restoration of tryptophan pyrrolase activity, and the similarity of the "transformations" induced by their DNA to those induced by v^+ DNA constitute the best evidence for this conclusion. It is unlikely that these alterations are simple, intermediate mutants. The transmissional and expressional uncertainties observed during the first few generations of the derivation of these stocks are unknown (contrary to some claims, the mutant v^2 is a null mutant).

In spite of their map position and functional properties, it is also unlikely that the alterations in "recessive transformants" are simple *su(s)* mutants. Again the evidence is strong, although not critical. It consists of the early uncertainties during the derivation of these stocks, of the differences in their recombinational properties from those of true *su(s)* mutants, of the differences in the ontogenetic development of their tryptophan pyrrolase activity from that of a *bona fide* *su(s)* mutant (Tobler, 1975), of the ability of their DNA to induce "transformations" like those induced by $su(s)^+ \ v^+$ DNA, and of the apparent transposition of their alteration from the locus of *su(s)* to that of *v*, where they behave exactly like those of "dominant transformants". Indeed, the last two observations strongly suggest that the alteration in "recessive transformants" consists of the introduced information of the v^+ gene at a heterologous position.

The question of models has been discussed in a previous review (Fox *et al.*, 1971b). There and elsewhere a model has been proposed which provides a conceptual framework for the complex and unconventional results reviewed above; it should be emphasized that it is regarded solely as heuristic in intent. The model suggests that *exosomes,* DNA segments of external origin, enter the cells of treated individuals and after an initial unstable period become associated with specific chromosome sites, homologous or heterologous. In contrast with the behavior of episomes in bacteria, no exchanges occur between exosome and chromosome. The exosome, therefore, is never integrated into the linear structure of the chromosome. It is held in association with its chromosomal site, replicates at least partially in step with that site, and is usually transmitted with it at the time of mitosis and meiosis. Occasional mistiming of exosome replication, or the loosening of its association with the chromosome, results in the loss of the exosome in some cell lineages or transposition to other chromosomal sites. At certain critical times, in some cell lineages, transcription of the DNA of the exosome and its associated chromosome segment is initiated. When the exosome is associated with its homologous chromosome site, this transcription is alternative - in any cell lineage either the DNA of the chromosome or the DNA of the exosome is transcribed. When the exosome is associated with a heterologous site, signals appropriate to its transcription are not present; instead, in some cell lineages it interferes with transcription of the associated chromosomal site. Individuals possessing a particular exosome in all of their cells, therefore, would always be phenotypic mosaics.

It will be noted that this model accounts for even the most unconventional of the results reviewed above, and reasons for prefering it over alternatives have been discussed (Fox *et al.*, 1971b). It implies mechanisms very different from those involved in bacterial transformation. It would be unprofitable to speculate further except to note that site recognition, either primary or following transposition, may be based on similarities between the nucleotide sequences in parts of the exosome and the chromosome sites with which it associates. It might also be suggested that the unusual chromosome anomalies observed at "transformed" sites reflect an unusual association of the introduced DNA with the chromosome.

FUTURE TASKS AND IMPLICATIONS

This review makes future tasks apparent. The most pressing problem is that of providing a *critical* demonstration that these effects result from gene transfer. Two systems are now available for such a demonstration. These involve the enzymes xanthine dehydrogenase and alcohol dehydrogenase and their respective structural genes. Work with the former has already been started, and work with the latter will be initiated in the near future.

The solution of this and other problems depends heavily on the development of methods capable of improving the yield of transformants. The recent development of methods for the injection of individual, well-staged eggs (Zalokar, 1971; Schubiger and Schneiderman, 1971; Illmensee, 1972; Okada, *et al.*, 1974) is already yielding promising results, as noted above, and may well contribute to this objective. It should also make possible such physical studies as dose-response relations, identification of the size-effective class of DNA, and studies of the early fate of introduced DNA.

Elucidation of the nature of the chromosomal anomalies at "transformed" sites is essential. Problems of transcription, however, may well be better studied in cell-culture systems.

The complexities encountered in this work serve as cautions against the expectation of simple results with other systems. In particular, the mosaic expression encountered here, if general, places obvious limitations on possible uses of gene transfer for gene therapy or in practical applications. It should be noted finally that this work with *Drosophila* has implications limited to the use of homologous DNA. The use of heterologous DNA may well involve an entirely different set of problems (Kleinhofs, this volume).

ACKNOWLEDGEMENTS

Supported by research grants from the National Institutes of Health (GM-11777 and GM15422). Laboratory of Genetics Paper No. 1869.

REFERENCES

Baglioni, C. (1960). Genetic control of tryptophan pyrollase in *Drosophila melanogaster* and D. *virilis. Heredity 15,* 87 - 96.

Baker, W. K. (1967). A clonal system of differential gene activity in *Drosophila. Devel. Biol. 16,* 1 - 17.

Beadle, G. W., and Ephrussi, B. (1937). Development of eye colors in *Drosophila:* Diffusible substances and their interrelations. *Genetics 22,* 76 - 86.

Becker, H. J. (1957). Uber Rontgenmosaikflecken und Defektmutationen am Auge von *Drosophila* und die Entwicklungsphysiologie des Auges. *Z. indukt. Abstam.- Verbungsl. 88,* 333 - 373.

Burch, J. W., and McBride, O. W. (1975). Human gene expression in rodent cells after uptake of isolated metaphase chromosomes. *Proc. Nat. Acad. Sci. (U.S.A.) 72,* 1797 - 1801.

Butenandt, A. (1953). Biochemie der Gene und Genwirkungen. *Naturwissenschaften 40,* 91 - 100.

Fincham, J. R. S., and Sastry, G. R. K. (1974). Controlling elements in maize. *Ann. Rev. Genetics 8,* 15 - 50.

Fox, A. S., and Valencia, J. I. (1975). Gene transfer in *Drosophila melanogaster:* Cytological alterations in the salivary chromosomes of transformed stocks. *Chromosoma 51,* 279 - 289.

Fox, A. S., and Yoon, S. B. (1966). Specific genetic effects of DNA in *Drosophila melanogaster. Genetics 53,* 897 - 911.

Fox, A. S., and Yoon, S. B. (1970). DNA-induced transformation in *Drosophila:* Locus-specificity and the establishment of transformed stocks. *Proc. Nat. Acad. Sci. (U.S.A.) 67,* 1608 - 1615.

Fox, A. S., Duggleby, W. F., Gelbart, W. M., and Yoon, S. B. (1970). DNA-induced transformation in *Drosophila:* Evidence for transmission without integration. *Proc. Nat. Acad. Sci. (U.S.A.) 67,* 1834 - 1838.

Fox, A. S., Yoon, S. B., and Gelbart, W. M. (1971a). DNA-induced transformation in *Drosophila:* Genetic analysis of transformed stocks. *Proc. Nat. Acad. Sci. (U.S.A.) 68,* 342 - 346.

Fox, A. S., Yoon, S. B., Duggleby, W. F., and Gelbart, W. M. (1971b). Genetic transformation in *Drosophila.* In *Informative Molecules in Biological Systems* (L. G. H. Ledoux, ed.), pp. 313-333. North-Holland Publishing Co., Amsterdam.

Fox, A. S., Parzen, S. D., Salverson, H., and Yoon, S. B. (1975). Gene Transfer in *Drosophila melanogaster:* Genetic transformations induced by the DNA of transformed stocks. *Genet. Res. 26,* 137 - 147.

Fox, A. S., and Allen, M. K. (1964). On the mechanism of deoxyribonucleate integration in Pneumoccocal transformation. *Proc. Nat. Acad. Sci. (U.S.A.) 52,* 412 - 419.

Friedman N. T., and Roblin, R. (1972). Gene therapy for human genetic disease? *Science 175,* 949 - 955.

Germeraad, S. E. (1975). Induction of genetic alterations in *Drosophila* by injection of DNA into embryos. *Genetics 80,* s34.

Hess, D. (1972). Transformationen and hoheren organismen. *Naturwissenschaften 59,* 348 - 355.

Illmensee, K. (1972). Developmental potencies of nuclei from cleavage, preblastoderm, and syncytial blastoderm transplanted into unfertilized eggs of *Drosophila melanogaster. Wilhelm Roux' Archiv 170,* 267 - 298.

Lees, A. D., and Waddington, C. H. (1942). The development of the bristles in normal and some mutant types of *Drosophila melanogaster. Proc. Roy. Soc. (London) B 131,* 87 - 110.

LeFevre, G., Jr. (1969). The eccentricity of vermilion deficiencies in *Drosophila melanogaster. Genetics 63,* 589 - 600.

Lefevre, G., Jr. (1971). Salivary chromosome bands and the frequency of crossing over in *Drosophila melanogaster. Genetics 67,* 497 - 513.

Lindsley, D. L., and Grell, E. H. (1968). *Genetic Variations of Drosophila melanogaster.* Carnegie Institution Publication No. 627, Washington, D. C.

McBride, W. O., and Ozer, H. L. (1973). Transfer of genetic information by purified metaphase chromosomes. *Proc. Nat. Acad. Sci. (U.S.A.) 70,* 1258 - 1262.

Mishra, N. D., and Tatum, E. L. (1973). Non-Mendelian inheritance of DNA-induced inositol independence in *Neurospora. Proc. Nat. Acad. Sci. (U.S.A.) 70,* 3875 - 3879.

Mishra, N. C., Niu, N. C., and Tatum, E. L. (1975). Induction by RNA of inositol independence in *Neurospora crassa. Proc. Nat. Acad. Sci. (U. S. A.) 72,* 642-645.

Nawa, S., and Yamada, M. A. (1968). Hereditary change in *Ephestia* after treatment with DNA. *Genetics 58,* 573 - 584.

Okada, M., Kleinman, I. A., and Schneiderman, H. A. (1974). Restoration of fertility in sterilized *Drosophila* eggs by transplantation of polar cytoplasm. *Devel. Biol. 37,* 43 - 54.

Schubiger, M., and Schneiderman, H. A. (1971). Nuclear transplantation in *Drosophila melanogaster. Nature 230,* 185 - 186.

Schwartz, A. G., Cook. P. R., and Harris, H. (1971). Correction of a genetic defect in a mammalian cell. *Nature, New Biology 230,* 5 - 8.

Sonnewblick, B. P. (1950). The early embryology of *Drosophila melanogaster.* In *Biology of Drosophila.* (M. Demerec, ed.), pp. 62-167. John Wiley & Sons, Inc., New York.

Stadler, D. R. (1973). The mechanism of intragenic recombination. *Ann Rev. Genetics 7,* 113 - 127.

Sturtevant, A. H. (1920). The vermilion gene and gynandromorphism. *Proc. Soc. Exp. Biol. Med. 17,* 70 - 71.

Tartof, K. D. (1969) Interacting gene systems: I. the regulation of tryptophan pyrrolase by the vermilion-suppressor of vermilion system in *Drosophila. Genetics 62,* 781 - 795.

Tobler, J. E. (1975) Dosage compensation and ontogenic expression of suppressed and transformed *vermilion* flies in *Drosophila. Biochem. Genetics 13,* 29 - 43.

Von Schaik, N., and Brink, R. A. (1959). Transpositions of modulator, a component of the variegated pericarp allele in maize. *Genetics 44,* 725 - 738.

Willecke, K., and Ruddle, F. H. (1975). Transfer of the human gene for hypoxanthine-guanine phosphoribosyl transferase via isolated human metaphase chromosomes into mouse L-cells. *Proc. Nat. Acad. Sci. (U.S.A.) 72,* 1792 - 1796.

Yoon, S. B., and Fox, A. S. (1965). Permeability of premature eggs from *Drosophila* collected with the ovitron. *Nature 206,* 910 - 913.

Zalokar, M. (1971). Transplantation of nuclei in *Drosophila melanogaster. Proc. Nat. Acad. Sci. (U.S.A.) 68,* 1539 - 1541.

Phage-Mediated Transgenosis in Plant Cells

Colin H. Doy

The substance of the presentation given in this meeting was similar to that delivered in Washington to the Workshop on Genetic Improvement of Seed Proteins organized by the National Science Foundation. It is not repeated here due to the commitment to publish in the Proceedings of that Workshop (Doy, 1975).

The following is a summary of the major points made about phage-mediated transgenosis. Transgenosis was discussed with emphasis on the fact that the term is not meant at this stage to represent a process synonomous with transduction or transformation but, instead, focuses attention that there is no claim for integration of foreign genes into the plant genome (nuclear or cytoplasmic) nor even inheritance.

Various changes in phenotype depend on the use of specialized transducing phage (λ or ϕ80) as vectors and occur only when relevant genes are present in the phage genome. New data were presented which suggest that phage-mediated transgenosis is eventually lethal. It was speculated that lethality related to transgenosis of genes whose products are of no advantage to the plant cells.

It is possible, although unlikely, that examples of balanced and selective transgenosis exist. Some cultures of *Lycopersicon esculentum* (tomato) cells originally presumed to be transgenosed with *lac* genes derived from *Escherichia coli* K12 have continued to grow on lactose medium indefinitely (at least three years of continuous sub-culture). These cultures never showed a high level of β–galactosidase activity and cannot be shown to contain an enzyme protected against heat denaturation by anti-serum to the pure *E. coli* enzyme. In these cultures, the only reason for assuming transgenosis is the biological response of growth which, unfortunately, can also occur for a few calluses not treated with phage. There remains a possibility that some or all of the phage-treated calluses represent examples of a balanced and selected gene transgenosis based on the synthesis of a hybird enzyme that cannot be protected by anti-serum to the *E. coli* enzyme. However a more cautious and less provocative interpretation is that growth on lactose is now due to an epigenetic or mutational event dependent only on the plant genome.

Data obtained in our laboratory (Doy *et al.,* 1972a, b; 1973a, b, c) and two others (Carlson, 1973; Johnson *et al.,* 1973) are consistent with the interpretation that the phenomenon of phage-mediated transgenosis is the consequence of transfer, transcription and translation of *E. coli* genes in higher plant cells. The phenotype is not permanent either because of lethality or lack of synchronized inheritance. This being so, methods of developing transgenosis further are suggested with special emphasis on the use of restriction enzymes to construct novel forms of DNA. It is suggested that the objectives should be two-fold; 1, limit the transgenote to the genes of relevance and 2, to integrate this into the existing nuclear or cytoplasmic genome of the recipient plant cell or whole plant. Other possibilities are that a means to ensure cytoplasmic inheritance of a plasmid transgenote might be developed.

Attention was drawn to the existing and potential biological dangers of transgenosis and it was recommended that plant scientists should participate in the general discussion of the possible hazards of genetic engineering and consider most carefully the implications of their own research.

Finally, although transgenosis was introduced as a term to emphasize a pheonotypic rather than a genetical consequence of gene transfer, it may in the long run prove to be a useful term to refer to a permanent genetical change mediated by a transgenote donated, at least in part, by a cell widely separated by evolution from the recipient so that a process of normal sexual mating is excluded. One can envisage reciprocity of transgenosis between procaryotes and eucaroytes which, in some examples, might be of practical benefit to mankind. One such goal might be the transfer of nitrogen fixation to the permanent genotype of important crop plants or, alternatively, by the creation of new symbiotic associations between plants and specially constructed microorganisms.

REFERENCES

Carlson, P.S. (1973). The use of protoplasts for genetic research. *Proc. Nat. Acad. Sci. (U.S.A.) 70,* 598 - 602.

Doy, C.H. (1976) Asexual approaches, including transgenosis and somatic cell hybridization, to the modification of plant genotypes and phenotypes. In *Genetic Improvement of Seed Proteins,*pp. 341-357, National Academy of Sciences, (U.S.A.) Washington.

Doy, C.H., Gresshoff, P.M., and Rolfe, B.G. (1972a). Transgenosis of the galactose operon from *Escherichia coli* to cultured cells of haploid *Lycopersicon esculentum* (tomato). *Proc. Aust. Biochem. Soc. 5,* 3.

Doy, C.H., Gresshoff, P.M., and Rolfe, B.G. (1972b). Transfer and expression (transgenosis) of bacterial genes in plant cells. *Search 3,* 447 - 448.

Doy, C.H., Gresshoff, P.M., and Rolfe, B.G. (1973a). Biological and molecular evidence for the transgenosis of genes from bacteria to plant cells. *Proc. Nat. Acad. Sci. (U.S.A.) 70,* 723 - 726.

Doy, C. H., Gresshoff, P.M. and Rolfe, B. G. (1973b). Transgenosis of bacterial genes from *Escherichia coli* to cultures of haploid *Lycopersicon esculentum* and haploid *Arabidopsis thaliana* plant cells. In *The Biochemistry of Gene Expression in Higher Organisms* (J.K.Pollak and J.Wilson Lee, eds.)p.21. Australia and New Zealand Book Co., Sydney.

Doy, C. H., Gresshoff, P.M., and Rolfe, B.G. (1973c). Time-course of phenotypic expression of *Escherichia coli* gene Z following transgenosis in haploid *Lycopersicon esculentum* cells. *Nature New Biology 244,* 90 - 91.

Johnson, C. B., Grierson, D., and Smith, H. (1973). Expression of λplac5 DNA in cultured cells of a higher plant. *Nature New Biology 244,* 105 - 106.

Physical and Biological Studies of DNA Uptake by Plants

A. Kleinhofs

Genetic engineering of cells and organisms has long been a dream of biologists working with eucaryotes. A few years back this dream appeared to be turning toward reality with reports that plant cells are capable of incorporating and replicating exogenously applied DNA. The remarkable studies of Ledoux and co-workers indicated that exogenously applied heterologous DNA could be taken up by a plant (Ledoux, 1965), integrated into (Ledoux and Huart, 1968; Ledoux *et al.*, 1971) and replicated along with the plant's genome (Stroun *et al.*, 1967; Ledoux and Huart, 1969; Ledoux *et al.*, 1972).

The evidence cited in the work of Ledoux is based primarily on identification of DNA species by their buoyant density in CsC1. Donor bacterial DNA of naturally high buoyant density was administered to plant seeds or shoots. Tritiated or ^{32}P labelled bacterial DNA was used for detection of uptake and integration. For detection of replication, unlabelled bacterial DNA was used, followed by a pulse of ^{3}H-thymidine. A crude DNA extract was prepared from various plant tissues and subjected to CsC1 density gradient centrifugation. A radioactive peak at the donor bacterial DNA density was taken as evidence for exogenous bacterial DNA uptake. A radioactive peak at a density intermediate between donor bacterial DNA and recipient plant DNA was taken as evidence for recombinant molecules. The intermediate density peak of DNA was further characterized by denaturation or fragmentation by sonication. Denaturation shifted the intermediate density peak as a single unit toward higher density. Fragmentation by sonication split the intermediate density peak of DNA into two density bands corresponding approximately to the donor and recipient DNA densities.

Ledoux's work has contributed tremendously to the stimulation of interest and imagination in the field of plant genetic engineering. Unfortunately, the prospects of genetic engineering with plants have suffered from the failure of other laboratories to verify the results of Ledoux and co-workers. Hotta and Stern (1971) reported that meiotic cells of lily, germinating barley embryos, and meristematic shoots of tomato take up DNA prepared from the bacterium, *M. lysodeikticus.* Part of the bacterial DNA was adsorbed to the nuclear membrane, but none of the bacterial DNA was present as such in the host nucleus. No evidence was found for integration of the bacterial DNA into the plant genome or its replication in the plant cells. Covalently linked molecular hybrids between bacterial and plant DNA were found only under abnormal physiological conditions induced by desiccation or by irradiation. These hybrid molecules were present either in the cytoplasm or at the surface of the nucleus but did not undergo replication. Bendich and Filner (1971) likewise were able to demonstrate the uptake of heterologous and homologous DNA by pea seedlings and tobacco cells in shake culture. No evidence for integration with the host DNA or replication of the exogenous DNA was found. Ohyama and co-workers (1972) and Heyn and Schilperoort (1973) reported the uptake of exogenous DNA by plant protoplasts. Heyn and Schilperoort (1973) found no evidence for integration of the exogenous DNA. Thus, there appears to be a general agreement in the literature about the uptake of exogenous DNA by plants, but there is no support for the integration and replication of the exogenous DNA in plant cells. More recent evidence indicates that the conclusions of integration and replication of exogenously applied DNA in plant cells may have been based on artifactual results (Kleinhofs, 1975; Kleinhofs *et al.*, 1975).

The failure to obtain confirming evidence on the integration and replication of exogenous DNA in plants prompted us to reexamine the phenomenon of exogenous DNA uptake by plant cells. The quantitative analysis of exogenous DNA uptake and the biological quality of the recovered DNA were of primary interest. We present here results on exogenous bacterial DNA uptake investigated by CsC1 density gradient analysis, DNA-DNA reassociation studies, and bacterial transformation assays. These techniques detect specific types of DNA molecules. CsC1 density gradient analysis is specific for the total DNA composition while DNA-DNA reassociation and bacterial transformation assays are specific for the bacterial donor DNA base sequences. The bacterial transformation assay detects only biologically active bacterial donor DNA molecules.

CsC1 DENSITY GRADIENT ANALYSIS

This technique is dependent upon the density of the DNA molecules. Since density is dependent upon the base compostion (GC content) or the presence of unusual bases, the technique is sensitive to the total composition of the DNA, but usually not to the base sequences present; the higher the GC content, the higher the CsC1 buoyant density. The administration of radioactively labelled bacterial DNA of a naturally high buoyant density to barley seeds, followed by isolation of total seedling root DNA and analysis in CsC1 gradients, always indicated the presence of the bacterial donor DNA (Fig. 1). The bacterial DNA appeared in the gradient as a high buoyant density radioactive peak. The size

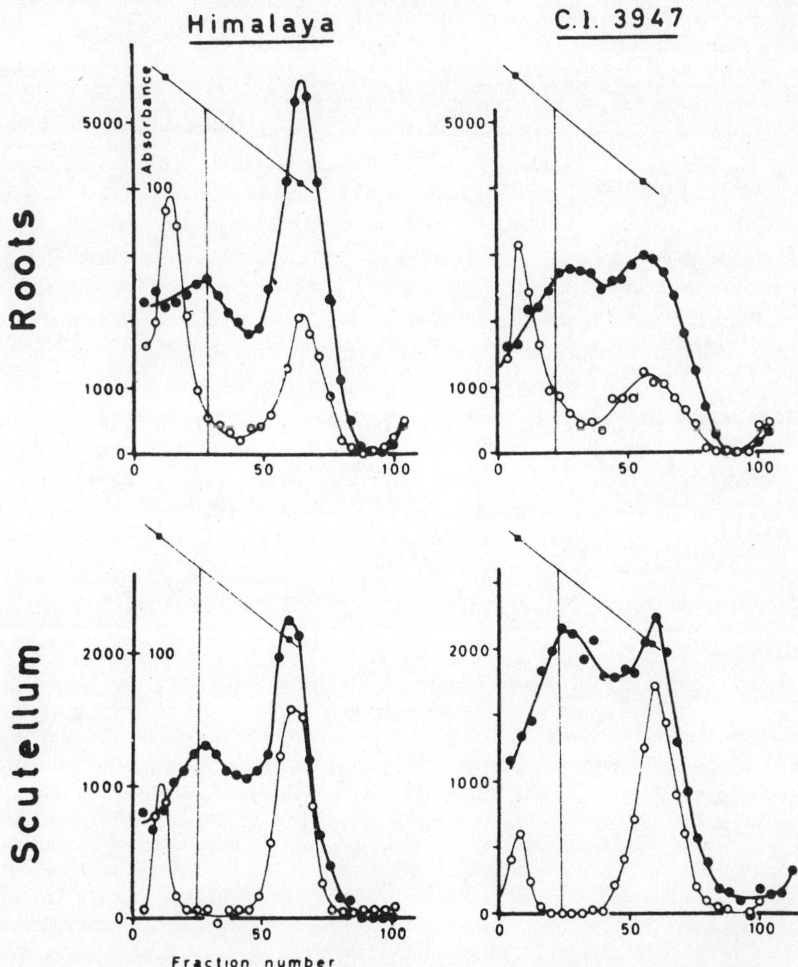

Fig. 1. CsCl density gradient analysis of barley root and scutellum DNA. Seeds were exposed to Sarcina flava ^{3}H-DNA (1.730 g/cm^3) for 12 hours, incubated in water for 36 hours and roots and scutellum excised. Absorbance peaks (open circles) are from phage 2C marker DNA (1.742 g/cm^3) and from barley DNA (1.702 g/cm^3). Radioactive peaks (closed circles) correspond with barley and Sarcina flava donor DNA densities. The labelling of the barley DNA is presumably from reutilization of S. flava ^{3}H-DNA breakdown products. The radioactive peak with S. flava density (indicated by thin vertical line) is from the donor DNA presumably taken up by the barley roots and scutellum.

TABLE 1
Quantitative estimates of exogenous donor bacterial DNA detected in barley seedlings by means of CsC1 density gradient analysis, DNA-DNA reassociation analysis and bacterial transformation assay.

Source of DNA Sample	Incub. time*	Donor bacterial DNA / Recip. barley DNA**
CsC1 density gradient analysis		
Himalaya roots	36 hrs.	2.4
Himalaya scutellum	36 hrs.	1.8
CI3947 roots	36 hrs.	3.0
CI3947 scutellum	36 hrs.	2.1
Proctor roots	72 hrs.	1.4
DNA-DNA reassociation analysis		
Himalaya, CI3947, or Ubamer seedlings	10 days	0.25
Bacterial transformation assay		
Proctor roots	2 days	0.14***

*Incubation time in water after termination of DNA treatment. **In ug/mg or bacterial genomes per haploid barley genome (using the approximation that the haploid barley genome is 10^3 times larger than the bacterial genome.) ***Maximum value obtained.

Treatment of barley seeds with exogenous donor bacterial DNA. Naked barley seeds, (cv. Himalaya, CI3947, and Ubamer) were sterilized in 5% calcium hypochlorite solution for 15 minutes. Covered barley seeds (cv. Proctor) were first treated with 50% (v/v) H_2SO_4 for 2 hours with gentle stirring at room temperature and extensively washed with distilled water and 10% Na_2CO_3 solution. This treatment removed most of the lemma and palea adhering to the caryopsis. Any remaining material was removed by hand. The clean seeds were then sterilized as for naked barley seeds, or in some cases the sterilization period was extended to 1 hour. The seeds were washed 5 times with sterile distilled water. All further operations were conducted under aseptic conditions. The seeds were soaked in distilled water for 11 hours, sectioned as described by Ledoux and Huart (1969), treated either with bacterial exogenous DNA in 0.01 M NaC1 or with 0.01 M NaC1.

The exogenous bacterial DNA treatments were conducted in ashing dishes or in 3.0 x 5.0 mm wells drilled in teflon discs. The treatment container containing the seeds was placed on water-saturated blotters in Petri dishes, sealed with tape, and left for 12 hours at room temperature in the dark. Afterwards the seeds were washed five times with distilled water. DNase treatments, when indicated, were with 10 ug/ml DNase (Sigma) for 10 minutes, followed by five washes with distilled water. The germinating seeds were transferred either to wells containing distilled water or perlite containing nutrient solution (Miksche and Brown, 1965) and allowed to grow for the indicated incubation time until harvested. Bacterial DNA was prepared and purified by CsC1 centrifugation as described by Lurquin and Behki (1975). Barley DNA was prepared by a modification (Kleinhofs, 1975) of a pronase procedure (Ledoux, 1969). Density gradient centrifugation was performed in 5 ml of CsC1 solution in a Spinco 40 rotor at 35,000 rpm for 60-64 hours. Fractions were collected from the bottom of the tube (Ledoux and Charles, 1972). DNA-DNA reassociation experiments were performed as described in detail previously (Kleinhofs, 1975). Bacterial transformation assays were performed as described in detail elsewhere (Ulrich, 1975).

and shape of the peak were variable, but it was usually small and broad, indicating that relatively small amounts of the donor DNA were present and that it had undergone some degradation. The quantity of bacterial donor DNA recovered per milligram of plant DNA could be estimated based on the area under the radioactive peak and the specific activity of the bacterial donor DNA (Table 1).

DNA–DNA REASSOCIATION ANALYSIS

DNA-DNA reassociation experiments provide a specific test for the presence of bacterial donor DNA base sequences. Thus, they are probably more reliable than CsC1 density gradient analysis. Considerably greater sensitivity can also be obtained with DNA-DNA reassociation analysis than with CsC1 density gradient analysis. In these experiments a very small quantity of radioactive bacterial DNA (the probe DNA) is reassociated in the presence of a large excess of unlabelled DNA from plants previously treated with exogenous bacterial DNA. The presence of bacterial DNA sequences in the plant DNA preparation accelerates the rate of reassociation of the bacterial DNA probe. The increase in the rate of probe reassociation is directly proportional to the quantity of bacterial DNA present in the plant DNA sample.

Unlabelled *E. coli* DNA was administered to barley seeds cv. Himalaya, CI3947 and Ubamer (See footnote to Table 1), followed by isolation of total seedling DNA after 10 days of growth. The purified plant DNA was divided into two samples. One sample was fragmented by shear forces, denatured, and used for DNA-DNA reassociation analysis with an *E. coli* ^{3}H-DNA probe. The results (Fig. 2) indicated the presence of a small quantity of about 0.25 *E. coli* genomes per haploid barley genome in the DNA preparations (Table 1).

The remaining DNA samples were used to test the integration or non-integration of the bacterial DNA with the plant DNA. The technique involves the renaturation of high molecular weight DNA at DNA concentrations and for time periods that only allow opportunity for reassociation of the repetitive plant DNA sequences (Varmus *et al.*, 1973). Since repetitive base sequences are interspersed with the unique base sequences, a complex DNA network is formed. This network can be collected by centrifugation. Any bacterial DNA that might have been integrated into the plant DNA would become a part of this network and co-sediment with it; any non-integrated bacterial DNA should remain in the

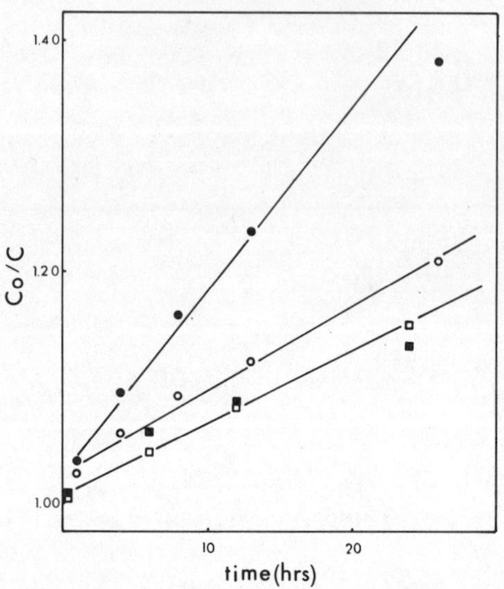

Fig. 2. DNA-DNA reassociation studies with barley seedling DNA. Seeds were exposed to E. coli DNA for 12 hours and then grown on perlite for 10 days. The purified seedling DNA was divided into two parts. One part of the DNA (circles) was fragmented by shear forces and reassociated at 2 mg/ml in the presence of E. coli ^{3}H-DNA probe (0.5 ug/ml). The filled circles represent average data from three different samples, each with a different cultivar (Himalaya, CI3947, and Ubamer. The slope of the line was identical for the three samples; for simplicity, only one value was plotted. The open circles represent data with DNA isolated from control barley plants. The second part of the seedling DNA sample (squares) was used for network formation (see text). The pellet DNA from all three cultivars was pooled, fragmented by shear forces, and reassociated as before. The filled squares represent data obtained with DNA from treated seedlings. The open squares represent data obtained with calf thymus control DNA. The slope of the plot is equal to kCo. Since k is constant for all E. coli probe DNA samples, any increase in the slope is due to an increased concentration of E. coli DNA. The plant DNA isolated from seedlings exposed to E. coli DNA shows the presence of E. coli sequences. After purification of the plant DNA by network formation this effect is lost, indicating the non-integrated nature of the bacterial DNA sequences detected in the total plant DNA preparation.

supernate. In control experiments where bacterial DNA was mixed with unlabelled barley DNA and networks were formed the bacterial DNA remained in the supernatant (Kleinhofs, 1975).

Hence, the DNA samples from barley seedlings were denatured and reassociated to produce these networks and these were pelleted. The DNA pellet was dissolved, fragmented by shear forces, and hybridized to an *E. coli* ^{3}H-DNA probe as before. *E. coli* DNA sequences were undetectable in this preparation (Fig. 2), indicating that the *E. coli* DNA sequences that were previously detected were not integrated into the barley genome.

BACTERIAL TRANSFORMATION ASSAYS

Another technique which is able to detect the possible presence of bacterial DNA and which is highly sensitive and base sequence specific is based on bacterial transformation. This technique has the additional advantage of detecting only biologically active bacterial donor DNA molecules.

Extensive reconstruction experiments showed that *B. subtilis* transforming DNA can be detected in the presence of a large excess of barley DNA and the activity approximately quantitated (Ulrich, 1975). Barley seeds were treated with *B. subtilis* transforming DNA and afterwards incubated in water for two days, the roots were excised and DNA prepared by a pronase method. The DNA samples obtained from these seedling roots contained some biological activity when tested in the *B. subtilis* transformation assay. The amount of activity was variable, depending upon such parameters as the competency of the *B. subtilis* cells and the amount of plant DNA added. By preparing a standard recovery curve, the amount of *B. subtilis* transforming DNA present in the barley DNA preparation could be estimated (Table 1). The maximum valve obtained was 0.14 bacterial genomes per haploid barley genome.

The presence of DNA able to cause *B. subtilis* transformation in barley seedling root DNA preparations raised the question of whether this DNA is inside the plant cells or merely trapped in intracellular spaces. Incubation of the seeds with DNase immediately after termination of the DNA treatment or just prior to root excision indicated (Fig. 3) that most, but not all, of the recoverable *B. subtilis* transforming DNA is destroyed. These results suggested that most, but not all, of the recovered DNA is extracellular. Interestingly, even though the two DNase treatments were separated in time by two days, the amount of DNA that was not digested was about the same. This suggests that no additional DNA entered the plant cells and became DNase resistant during the two day period after DNA treatment.

DISCUSSION

The experiments that have been described were designed to determine whether exogenous bacterial DNA could be detected in DNA preparations obtained from plants grown from seeds exposed to exogenous bacterial DNA. Three different techniques were utilized: CsC1 density gradient analysis, DNA-DNA reassociation analysis, and bacterial transformation assays. Positive results

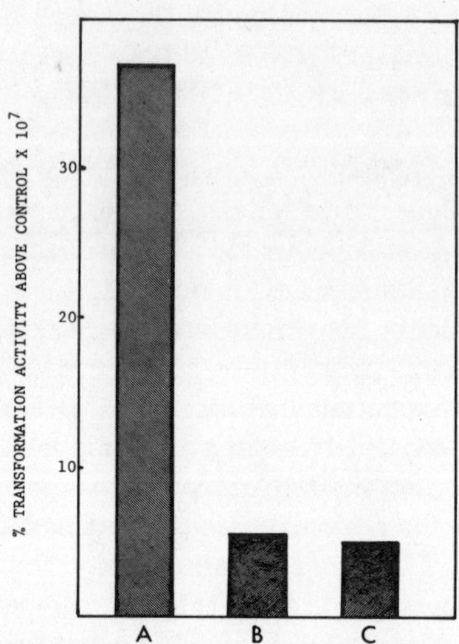

Fig. 3. B. subtilis transformation assay of barley root DNA. Seeds were exposed to B. subtilis DNA for 12 hours, incubated in water for 2 days and the roots were excised and their DNA isolated. Samples B and C were treated with DNase (10 ug/ml, 10 min.) either immediately after incubation in the B. subtilis DNA (B) or just prior to excision of the roots (C). The total DNA concentration in the transformation mixture was 1.9 ug/ml. DNase treatment (Samples B and C) resulted in the loss of about 85% of the recoverable transformation activity. The DNase resistant portion is the same at both times of DNase treatment.

were obtained with all three techniques. Results with the DNA-DNA reassociation and bacterial transformation techniques are particularly important because these techniques are base sequence specific and are thus less likely to be subject to artifacts.

Quantitative estimates of the amount of exogenous bacterial DNA detected by the various techniques indicated that small but significant amounts of bacterial DNA were detected in barley seedlings even 10 days after treatment with the exogenous bacterial DNA. Other investigators have reported that exogenous DNA can exist in plants for long periods of time (Ledoux *et al.*, 1971; Kado and Mojica-a, 1975). The quantity of DNA estimated by the DNA-DNA reassociation technique was approximately 0.25 *E. coli* genomes per haploid barley genome. Obviously, this is a large amount of genetic information if the exogenous DNA is stably maintained by the plant. The amount of exogenous bacterial DNA estimated by the bacterial transformation assay was an order of magnitude lower than the amount estimated by the CsC1 density gradient technique in comparable experiments. This indicated that most of the exogenous bacterial DNA co-isolated with the plant DNA is no longer biologically active as tested by the bacterial transformation assay. This may be due to partial degradation of the exogenous DNA. Morrison and Guild (1972) have shown that the lower limit molecular weight of transforming DNA for *B. subtilis* is about 1.2×10^6.

In calculating the amount of potentially functional exogenous bacterial DNA present in the plant DNA preparation, consideration must be given to the observation that only about 15% of the DNA found by the bacterial transformation assay was resistant to DNase digestion administered after DNA treatment. Thus, actually only 15% of 0.14 ug or about 0.02 ug exogenous bacterial DNA per milligram of host plant DNA, existed in potentially functional form and possibly inside the plant cells. This estimate itself is probably high, since it was made only two days after termination of the treatment with the exogenous bacterial DNA. We know that longer incubation of barley seedlings in water after DNA treatment decreases the amount of exogenous bacterial DNA to levels not reliably detected by the bacterial transformation assay (Ulrich, 1975).

The haploid barley genome size is estimated to be 3.5×10^{12} daltons, or 5×10^9 nucleotide pairs (Sciaky, 1973). The *E. coli* genome size is 2.9×10^9 daltons or 4.5×10^6 nucleotide pairs. Since the haploid barley genome is approximately 1,000 times larger than the *E. coli* genome, the estimate of micrograms of exogenous bacterial DNA per milligram of host barley DNA also approximates the number of *E. coli* genomes per haploid barley genome (Table 1).

Taking the above facts in consideration, one must conclude that the amount of exogenous bacterial DNA apparently taken up by the plant cells could contain a large amount of genetic information. Thus, it becomes paramount to determine if this DNA is really present inside the cells and if it is maintained by the plant through the sexual cycle.

The question of exogenous DNA penetration inside the cell membrane or adsorption to the cell wall or membrane has been considered by only a few investigators. It is generally assumed that exogenous DNA co-isolated with plant DNA after washing or DNase treatment of the plant organs, cells or protoplasts is inside the plant cell membrane. Recently, Lurquin and Behki (1975) have questioned this assumption and presented data from studies with *Chlamydomonas* that suggest that no detectable exogenous DNA penetrates the plant cell membrane. Thus, it seems possible that uptake of exogenous DNA by plant cells represents persistant exogenous DNA bound to plant cell walls and membranes in a DNase resistant form.

The only unambiguous resolution of this question would be to demonstrate replication of the exogenous DNA or the presence of a new function in the plant cells that can be directly traced to the exogenous DNA genetic information. In extensive experiments, Kleinhofs (1975) and Kleinhofs and co-workers (1975) were unable to demonstrate replication of the exogenous DNA in barley, pea, or tomato plant systems. Others (Kado and Lurquin, 1975; Lurquin and Hotta, 1975) have also reported negative results in their attempts to detect replication of exogenous DNA in *Phaseolus* seedlings and *Arabidopsis* cells in culture. Thus, outside of the reports by Ledoux and co-workers, no evidence for replication of exogenous DNA in plants exists.

The presence of a renewed function, presumably due to exogenous DNA in *Arabidopsis,* has been reported (Ledoux *et al.*, 1974). These results are very interesting but must be interpreted with caution.

SUMMARY

Evidence presented here suggests that exogenous DNA is taken up by plant cells. Quantitative estimates of the exogenous DNA present in plant cells indicates the possibility for addition of considerable new genetic information. The experiments reported here, and those available in the literature, do not differentiate between DNA actually inside the plant cell and DNA adsorbed in a DNase resistant form to plant cell walls or membranes. If the exogenous DNA detected in these experiments is inside the cell, then there is some hope that some of this genetic information may be useful to the plant and may become stably maintained within the plant cells. On the other hand, if the exogenous DNA is nonspecifically adsorbed to the cell walls and membranes, then techniques for introducing the exogenous DNA within the plant cell need to be developed before further progress in plant genetic engineering can be achieved.

ACKNOWLEDGEMENTS

Scientific Paper No. 4465. College of Agriculture Research Center, Washington State University, Pullman, Project 1920. Supported in part by National Institutes of Health Grant No. ESO1009.

REFERENCES

Bendich, A. J., and Filner, P. (1971). Uptake of exogenous DNA by pea seedlings and tobacco cells. *Mutation Res. 13,* 199 - 214.

Heyn, R. F., and Schilperoort, R. A. (1973). The use of protoplasts to follow the fate of *Agrobacterium tumefaciens* DNA on incubation with tobacco cells. *Colloques Internationaux C.N.R.S., 212,* 385 - 395.

Hotta, Y., and Stern, H. (1971). Uptake and distribution of heterologous DNA in living cells. In *Informative Molecules in Biological Systems* (L. Ledoux, ed.), pp. 176 - 186. American Elsevier Publ. Co., Inc., New York.

Kado, C. I., and Lurquin, P. F. (1975). Studies on *Agrobacterium tumefaciens* IV. Non-replication of the bacterial DNA in mung bean *(Phaseolus aureus). Biochem. Biophys. Res. Comm. 64,* 175 - 183.

Kado, C. I., and Mojica-a, T. (1975). Intercellular persistence of *Bacillus subtilis* transforming DNA in *Phaseolus aureus* plants. *Arch. Intern. Physiol. Biochim. 83,* 190.

Kleinhofs, A. (1975). DNA-hybridization studies of the fate of bacterial DNA in plants. In *Genetic Manipulations with Plant Materials* (L. Ledoux, ed.), Plenum Press, New York.

Kleinhofs, A., Eden, F. C., Chilton, M - D., and Bendich, A. J. (1975). On the question of the integration of exogenous bacterial DNA into plant DNA. *Proc. Natl. Acad. Sci. (U.S.A.) 72,* 2748 - 2752.

Ledoux, L. (1965). Uptake of DNA by living cells. In *Progress in Nucleic Acid Research and Molecular Biology* (J. N. Davidson and W. Cohn, eds.), Vol. 4, pp. 231 - 267. Academic Press, New York.

Ledoux, L., and Huart, R. (1968). Integration and replication of DNA of *M. lysodeikticus* in DNA of germinating barley. *Nature 218,* 1256 - 1259.

Ledoux, L., and Huart, R. (1969). Fate of exogenous bacterial deoxyribonucleic acids in barley seedlings. *J. Mol. Biol. 43,* 243 - 262.

Ledoux, L., Huart, R., and Jacobs, M. (1971). Fate of exogenous DNA in *Arabidopsis thaliana.* Translocation and integration. *Eur. J. Biochem. 23,* 96 - 108.

Ledoux, L., and Charles, P. (1972). On the use of preparative CsCl gradients. In *Uptake of Informative Molecules by Living Cells* (L. Ledoux, ed.), pp. 29-46. North-Holland Pub. Co., Amsterdam.

Ledoux, L., Huart, R., and Jacobs, M. (1972). Fate and biological effects of exogenous DNA in *Arabidopsis thaliana.* In *The Way Ahead in Plant Breeding* (F. G. H. Lupton, G. Jenkins and R. Johnson, eds.), pp. 165 - 184. Eucarpia, Cambridge.

Ledoux, L., Huart, R., and Jacobs, M. (1974). DNA-mediated genetic correction of thiamineless *Arabidopsis thaliana. Nature 249,* 17 - 21.

Lurquin, P. F., and Behki, R. M. (1975). Uptake of bacterial DNA by *Chlamydomonas reinhardi. Mutation Res. 29,* 35 - 51.

Lurquin, P. F., and Hotta, Y. (1975). Reutilization of bacterial DNA by *Arabidopsis thaliana* cells in tissue culture. *Plant Science Letters 5,* 103 - 112.

Miksche, J. P., and Brown, J. A. M. (1965). Development of vegetative and floral meristems of *Arabidopsis thaliana. Amer. J. Bot. 52,* 533 - 537.

Morrison, D. A., and Guild, W. R. (1972). Activity of deoxyribonucleic acid fragments of defined size in *Bacillus subtilis* transformation. *J. Bacterial. 112,* 220 - 223.

Ohyama, K., Gamborg, O. L., and Miller, R. A. (1972). Uptake of exogenous DNA by plant protoplasts. *Can. J. Bot. 50,* 2077 - 2080.

Sciaky, D. (1973). Characterization of isolated nuclei from ungerminated barley embryos and uptake of bacterial DNA by these nuclei. M. S. Thesis, Washington State University, Pullman, Washington.

Stroun, M., Anker, P., and Ledoux, L. (1967). DNA replication in *Solanum lycopersicum* esc. after absorption of bacterial DNA. *Currents in Mod. Biol. 1,* 231 - 234.

Ulrich, T. H. (1975). Recovery of *Bacillus subtilis* transforming DNA taken up by barley *(Hordeum vulgare* L.*)* seedlings. M. S. Thesis, Washington State University, Pullman, Washington.

Varmus, H. E., Vogt, P. K., and Bishop, J. M. (1973). Integration of deoxyribonucleic acid specific for Rous sarcoma virus after infection of permissive and non-permissive hosts. *Proc. Natl. Acad. Sci. (U.S.A.) 70,* 3067 - 3071.

Molecular Genetic Modification of Legumes

F. B. Holl

". . . I would urge botanists to tackle the problem of nitrogen deficiencies in the living world with boldness. We need to devise more efficient ecosystems, put root nodules on cereal crops and put nitrogen fixing chloroplasts in their leaves. . .", so spoke J. van Overbeak at the Symposium on World Food Supply at the XIth International Botanical Congress in 1969. Have we progressed toward this goal in subsequent years?

Biological fixation of dinitrogen provides a means of maintaining existing levels of bound nitrogen at agricultural sites where they are required for primary production. Much stress has been placed on soil nitrogen reserves by the increased requirements of the rapidly expanding world population. Our society has turned largely to the use of nitrogen fertilizers to provide the necessary combined nitrogen. Such use has been advocated in part, on the fallacious assumption that the energy cost to the plant in fixing dinitrogen could be conserved if fixation is carried out chemically in a factory. With the report that the nodulated root has an efficiency level similar to that of an uninoculated root in assimilating nitrate (Minchin and Pate, 1973), coupled with rapidly rising cost of fertilizar production, it appears more advisable to allow legumes to utilize solar energy for the fixation of dinitrogen (or to exploit other sources of dinitrogen fixation such as blue-green algae in rice crops). Dinitrogen fixation whether by legumes or in other forms will not be the answer for all our agricultural problems but it does represent a significant force for large-scale improvement.

Manipulations of symbiotic dinitrogen fixation have been directed mainly toward selection of effective bacterial strains. However it now appears that alterations in inoculum strain do not offer the potential for major increases in fixation. The regulation of dinitrogen fixation by host plant genes has not received similar attention (Holl and LaRue, 1975). Before we attempt radical changes in molecular genetic modification between species, it is essential to have a thorough understanding of the genetic basis of the characters to be exploited. Furthermore, the first stages in such a process would be to examine such modifications within plant genera using a system with simple, defined genetic differences. Such a system may be the symbiotic dinitrogen fixation process in field peas.

BIOCHEMICAL AND GENETIC ASPECTS OF SYMBIOTIC DINITROGEN FIXATION IN LEGUMES

Symbiotic dinitrogen fixation in legumes is a complex interaction at three levels: (1) between two organisms (plant and bacterial) and the environment, (2) between two genomes (plant and bacterial), and (3) between two physiological systems (photosynthesis [carbon fixation] and dinitrogen fixation). An appreciation of this complexity is necessary to understand the potential for modification. The biochemical stages of symbiotic dinitrogen fixation are summarized in Figure 1. The specificity of the response may be considered at a number of different levels of interaction.

(1) Colonization of root surface - Rhizobial growth appears to be more stimulated in the root zone of legumes than of grasses (Vincent, 1974).

(2) Host root-hair response - In many instances, bacterial invasion takes place via the root hairs of the host. These hairs show a specific curling response to rhizobia in the rhizosphere (Yao and Vincent, 1969).

(3) Host invasion - The actual process of invasion is a specific interaction affected by the genetics of both partners. Hosts vary widely in their susceptibility to infection and in the range of rhizobial types to which they are susceptible.

(4) Nodule development and maturation - Following invasion, the situation may range from complete inability to develop additional nodule material to proper nodule development with maximum dinitrogen fixation capacity. Both host and bacterium contribute to the range of types which may be encountered. Dinitrogen fixation cannot occur without the establishment and maintenance of an efficient, integrated metabolism between the macrosymbiont and the microsymbiont.

Fixation of atmospheric dinitrogen is catalyzed by the enzyme nitrogenase, a complex molybdenum iron-protein. The ability of nitrogenase to reduce a number of alternative substrates such as acetylene has been used to develop a useful assay system (Hardy *et al.,* 1968). This system has been modified to provide a non-destructive assay of enzyme activity in intact plant material (LaRue and Kurz, 1973).

Nitrogenase activity is dependent upon the proper functioning of a number of ancillary processes in addition to the structural development of the nodule. The energy requirement (as adenosine triphosphate, ATP) and the reductant necessary to supply electrons are both obtained via photosynthates translocated from the top of the plant. This relationship emphasizes the complex interaction at the system level between photosynthesis and dinitrogen fixation. In addition, nitrogenase is a very oxygen-sensitive enzyme. To maintain the required low oxygen tension in the nodule, large amounts of leghaemoglobin are present,

Fig. 1. Root nodule development in legumes

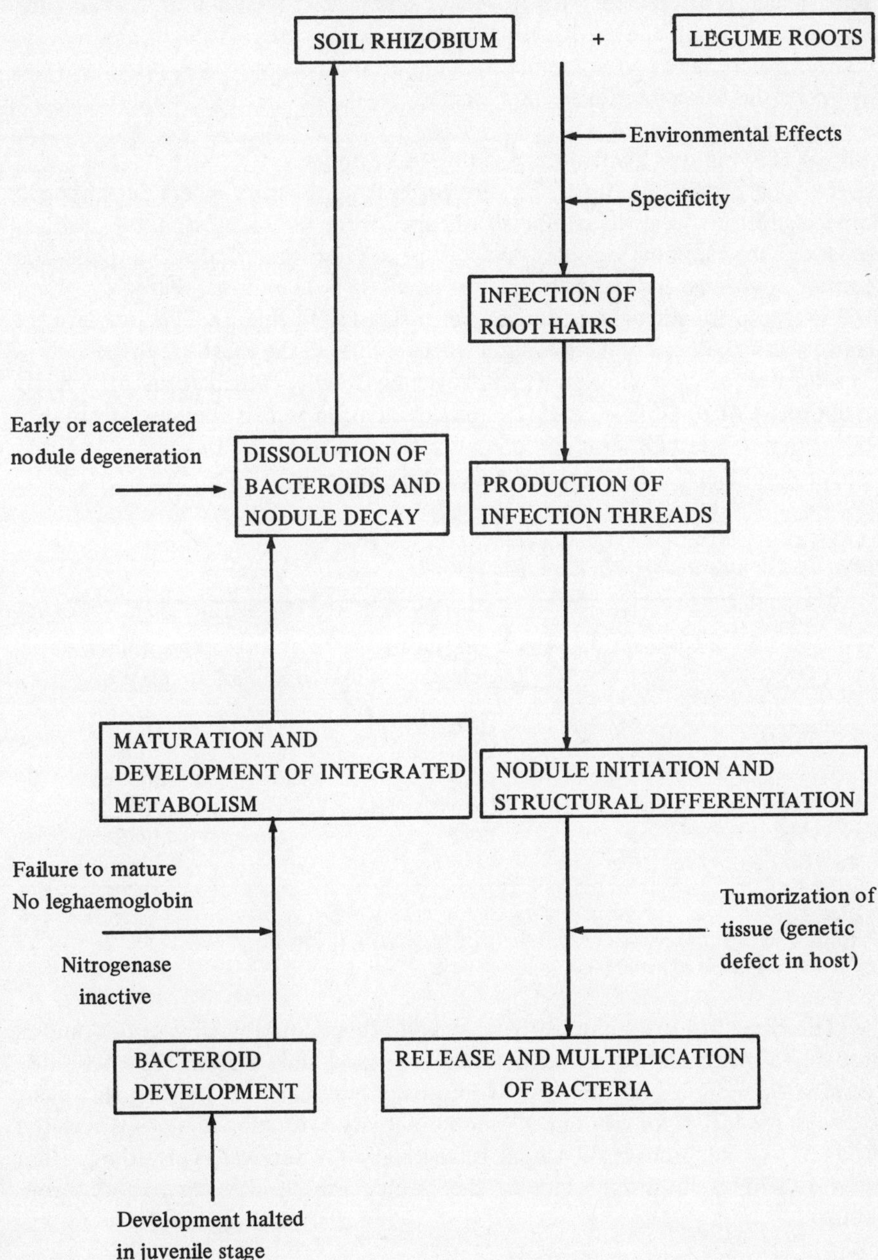

functioning as an oxygen scavenger. Atmospheric nitrogen is reduced to ammonia which is a repressor of nitrogenase synthesis. Utilization of the ammonia end product is an additional requirement for the proper functioning of the fixation process and further emphasizes the intimate balance that exists between the upper and lower portions of the plant.

Control of Symbiotic Dinitrogen Fixation in Legumes

In spite of their ability to fix dinitrogen from the atmosphere, legumes still derive significant levels of combined nitrogen from the soil. Estimates of nitrogen fixed in common legumes such as peas (Kurz and LaRue, unpublished results), soybeans and peanuts (Hardy *et al,* 1968), indicate that only about 30% of the plant nitrogen is derived from fixation (Table 1). This low level of fixation and the requirement for soil nitrogen are, in the most simplistic view, a consequence of two factors: (1) nitrogenase activity which turns off rapidly at the onset of pod-filling, and (2) both nodulation and fixation are sensitive to soil nitrogen levels. Other environmental factors also can be important - soil pH, temperature, light (Lie, 1974) and water stress (Sprent, 1971).

TABLE 1
Nitrogen accumulation by nodulated field peas.*

Variety or Line	Plant N at maturity (mg/plant)	Plant N fixed (mg/plant)
Trapper	1070	210 (20%)
MP693	1980	350 (18%)
MP766	1550	440 (29%)
340	1270	190 (15%)
7608	1270	350 (28%)
109	250	90 (36%)

**Total plant N and nitrogenase were determined weekly on field-grown peas. Nitrogen fixation over the season was estimated from summation of nitrogenase activities (LaRue and Kurz, unpublished results).*

The drop in nitrogenase activity at pod-filling is particularly critical and is probably a manifestation of the competitive source-sink relationship established between the nodule and the seed. If photosynthate becomes limiting, the plant balances the need for continued nodule activity with the need to divert the product to seed production which is necessary for survival. This competition again underlines the importance of the photosynthesis-nitrogen fixation interaction.

These aspects of the control of symbiotic dinitrogen fixation in legumes suggest two obvious questions which bear on the potential for molecular modification of legumes via DNA–feeding: (1) can symbiotic dinitrogen fixation in legumes be improved and (2) can a symbiotic-type of dinitrogen fixation be incorporated into other crop plants such as wheat and corn?

DNA Feeding To Plants

Much publicity has attended the suggestion that DNA-feeding might be used to transfer genetic information across the natural barriers between plant genera. Presented in Table 2 is a summary of experiments involving plant material and DNA-feeding. A variety of techniques and systems have been utilized, with reports of both successes and failures. The work of Ledoux and co-workers (Holl et al., 1974) represents the most comprehensive and rigorous biological and biochemical examination of DNA feeding. Their work has come under criticism from a number of sources (Hotta and Stern, 1971; Bendich and Filner, 1971; Johnson and Grierson, 1974).

MOLECULAR MODIFICATION OF DINITROGEN FIXATION IN LEGUMES

We are analyzing the biochemical genetics of dinitrogen fixation in field peas and evaluating this material as a model system for genetic information transfer via isolated DNA. The work has four main areas directed at some of the problems discussed in the previous sections of this paper: (1) marker systems,(2) DNA isolation procedures, (3) DNA-feeding methods, and (4) selection systems.

Marker Systems

Nodulation and dinitrogen fixation are easily assayed biological markers. We have been developing a donor/recipient pair of near isogenic lines of field pea differing in a single gene involved in symbiotic dinitrogen fixation. Our experiments to date, however, have utilized PRL-H722 as the recipient line for DNA isolated from Trapper, a normal small-seeded field pea variety (see Figure 2). PRL-H722 is a non-nodulating, non-fixing mutant grown from seed originally obtained from Dr. T. A. Lie of Wageningen, The Netherlands. The mutant has two independent host genes which appear to control nodulation and fixation respectively (Holl and LaRue, 1975; Holl, 1975a).

DNA Isolation

A rapid, reproducible technique has been developed for DNA isolation from seedlings (Holl *et al.*, 1974, Holl, 1975b). We are trying to determine whether or not a correlation exists between a number of physical and chemical characteristics of the isolated DNA and the ability of exogenous DNA to induce changes in the recipient material.

TABLE 2
DNA-feeding experiments in higher plants.

Donor	Recipient	Marker/Results	Reference
Bacterial DNA	Barley seeds	Radioisotope/ Uptake, integration	Ledoux and Huart (1969)
Bacterial DNA	Arabidopsis seeds	Radioistope, thiamine auxotrophy/ Uptake, translocation, mutant repair	Ledoux *et al.* (1974)
Bacterial DNA	Tomato apical shoots	Radioisotope/ Uptake, replication	Stroun *et al.* (1967)
Bacterial DNA	Lily cells Barley seedlings	Radioisotope/ Uptake poor	Hotta and Stern (1971)
Petunia DNA	Petunia seedlings	Flower color/ Mutant repair, leaf shape altered	Hess (1972)
Tobacco DNA	Tobacco	Hypoxanthine requirement/ Uptake, no repair	Carlson (1972)
Bacterial DNA	Plant protoplasts	Radioisotope/ Uptake	Ohyama *et al.* (1972)
Bacteriophage (lac+)	Tomato callus (lac-)	Lactose auxotrophy/ Growth on lactose, immunological cross reaction	Doy *et al.* (1973)
Bacteriophage (lac+)	Sycamore cell suspension	Lactose auxotrophy/ Growth on lactose	Johnson *et al.* (1973)
Bacterial DNA	Pea seedlings Tomato cells	Radioisotope/ Uptake poor	Bendich and Filner (1973)
Petunia DNA	Petunia protoplasts	Radioisotope/ Uptake	Hoffman(1973)
Bacterial DNA	Soybean protoplasts	Lactose and mannitol auxotrophy/ Growth on lactose and mannitol	Holl *et al.* (1974)
Field pea DNA	Field pea seeds	Lack of nodules/ Nodulation	Holl *et al.* (1974)

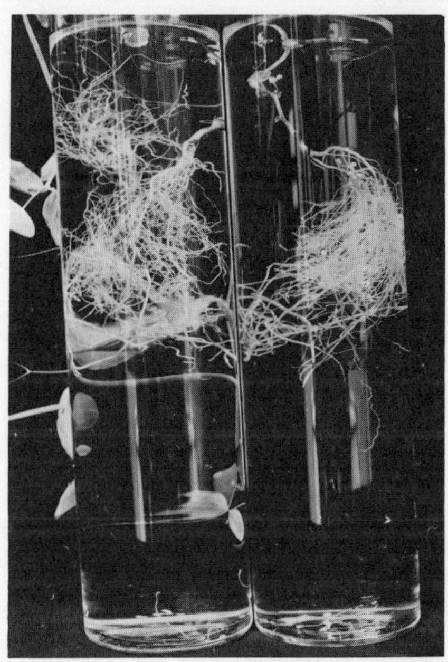

Fig 2. Nodulation and dinitrogen fixation in field peas. Wild-type and mutant lines grown in sand and vermiculite with nitrogen-free medium and inoculated with a mixed commercial culture of Rhizobium leguminosarum. a. PRL-H722, mutant non-nodulating, non-fixing line b. Trapper, wild-type nodulating, fixing line c. Root system of PRL-H722 (no nodules) d. Root system of trapper (nodulated)

DNA-feeding Systems

Little effort has been made to examine the phenomenon of "competence" in studies of DNA feeding in higher plant material or to evaluate the advantages of various donor/recipient systems such as seeds, plants, protoplasts, isolated DNA and bacteriophages. Although we have limited our donor material to isolated DNA, a number of our experiments have been designed to utilize different recipient material and to determine the possible advantages and disadvantages with reference to their use and potential.

Selection Systems

Significant progress likely will not be made in molecular genetic modification until suitable marker systems which lend themselves to selective pressures have been examined rigorously.

EXPERIMENTAL RESULTS

DNA Preparations

Our preliminary results on the characteristics of plant DNA preparations for use in feeding experiments are in agreement with those reported for bacterial DNA by Ledoux and co-workers (1972). We avoid the use of chloroform or phenol deproteinizations and never dry the preparations with ether or lyophilization. Clean, high molecular weight material with ultraviolet spectral characteristics 258/235 ≥ 1.8 and 258/280 ≥ 1.8 are most desirable. We are attempting to correlate biological activity with such physico-chemical characteristics as U. V. spectrum, melting profile and sedimentation properties although Ledoux and co-workers (1972) have indicated that such a correlation may not necessarily exist or may be difficult to assess.

DNA-feeding Systems

DNA-feeding experiments have been carried out in our laboratories with protoplasts, seeds and shoot tips (Ohyama *et al.*, 1972; Holl *et al.*, 1974). Varying degrees of success have been obtained in our own experiments and those reported in the literature. DNA quality may be one of several contributing factors. Variability may be due, in part, to the relative "competence" level of the recipient material. In particular, deoxyribonuclease (DNase) activity may play an important role in the uptake and initial fate of the fed DNA. DNase activity has been a problem in some experiments (Bendich and Filner, 1971), although Ledoux and co-workers suggested that degradation did not occur in experiments with intact seeds. Low levels of DNase activity occurred in some protoplasts and in intact seed within the 24 to 48 hours of imbibition and germination (Holl, 1973). We also have been able to detect little or no DNase activity in pea shoot tips.

The absence of DNase activity may reflect the receptivity of the material to exogenous DNA. Seeds, shoot tips and protoplasts each have particular merits as recipient systems. Seeds and shoot tips develop into plants which are readily examined and assayed, and which produce seeds for analysis of subsequent generations. Shoot tips may be superior to seeds because substantially less DNA is required to treat a given plant population. In terms of population size and DNA requirements, protoplasts remain the material of choice. However, their use for the present is seriously limited by the lack of procedures for inducing cell wall regeneration, cell division, and plant production.

DNA-feeding Experiments With Seeds

The majority of our experiments have utilized germinating seeds as the recipient material. DNA isolated from wild type, nodulating, N-fixing field peas was fed to a mutant line unable to nodulate or fix dinitrogen. After sterilization in ethanol and Javex[R], seeds were incubated for 96 hours at 28°C in the dark in a droplet (100-200 μl) of DNA (0.5 to 1 mg/ml) on water agar. During this time, most of the seeds germinated to produce roots and shoots. The seedlings were then transferred to a mixture of equal volumes of sand and vermiculite and inoculated with a commercial rhizobium inoculum. After approximately four weeks growth in nutrient solution lacking nitrogen, plants were examined for nodulation and acetylene reduction. Two "repaired" nodulated plants were found among 409 progeny from seed of non-nodulating field peas treated with DNA from nodulating field peas. No nodulating plants occurred among 145 from seed treated only with SSC (sodium saline citrate) or among 238 progeny from calf-thymus-DNA treated seed. It should emphasized that genetic analysis of the "repaired" progeny is essential.

I have indicated the complexity of the symbiotic dinitrogen-fixing relationships between plant and bacteria and within the plant to emphasize the difficulties that will have to be overcome in any attempt to transfer dinitrogen fixing ability to a non-fixing plant species. The potential for such an exchange and whether the complex symbiotic type of fixation is the best mode to examine will depend upon a comprehensive knowledge of the genetics of dinitrogen fixation and an exhaustive evaluation of the DNA-feeding technique in a genetically characterized system.

AKNOWLEDGEMENTS

The technical assistance of Mr. D. Olson is gratefully acknowledged.

REFERENCES

Bendich, A.J., and Filner, P. (1971). Uptake of exogenous DNA by pea seedlings and tobacco cells. *Mut. Res. 13*, 199-214.

Carlson, P.S. (1972). Attempts to detect DNA mediated transformation in a higher plant. *Genetics 71*, 59.

Doy, C.H., Gresshoff, P.M., and Rolfe, B.G. (1973). Biological and molecular evidence for the transgenosis of genes from bacteria to plant cells. *Proc. Nat. Acad. Sci. U.S.A. 70*, 723-26.

Hardy, R.W.F., Holsten, R.D., Jackson, E.K., and Burns, R.C. (1968). The acetylene-ethylene assay for N_2 fixation: Laboratory and field evaluation. *Plant Physiol. 43*, 1185-1207.

Hess, D. (1972). Transformation an hoheren Organismen. *Naturwissen. 59*, 348-355.

Hoffman, F. (1973). Die Aufnahme doppelt-markerter DNS in isolierte Protoplasten von *Petunia hybrida. Z. Pflanzenphysiol. 69*, 249-261.

Holl, F.B. (1973). Cellular environment and the transfer of genetic information. *Colloq. Int. Cent. Nat. Rech. Sci. 212*, 509-516.

Holl, F. B. (1975a). Host plant control of the inheritance of dinitrogen fixation in the *Pisum-Rhizobium* symbiosis. *Euphytica. 24,* 767-770.

Holl, F. B. (1975b). DNA isolation from plants for use in DNA feeding experiments. In *Plant Tissue Culture Methods.* O. L. Gamborg and L. R. Wetter, eds.), pp. 65-69. National Research Council of Canada.

Holl, F. B., Gamborg, O. L., Ohyama, K., and Pelcher, L. (1974). Genetic transformation in plants. In *Tissue Culture and Plant Science* (H. E. Street, ed.) pp. 301-327. Academic Press, New York.

Holl, F. B., and LaRue, T. A. G. (1975). Genetics of Legume Plant Hosts. In *Proc. of the International Symposium on Nitrogen Fixation.* Pullman Washington. (In Press).

Hotta, Y., and Stern, H. (1971). Uptake and distribution of heterologous DNA in living cells. In *Informative Molecules in Biological Systems:* (L. Ledoux, ed.), pp. 176-186. North-Holland Publishing Co., Amsterdam.

Johnson C. B., Grierson, D., and Smith, H. (1973). Expression of plac 5 DNA in cultured cells of a higher plant. *Nature New Biology 244,* 105-107.

Johnson, C. B., and Grierson, D. (1974). The uptake and expression of DNA by plants. Commentaries in Plant Science No. 9 In *Current Advances in Plant Science.* (H. Smith, ed.), pp. 1-12. Robert Maxwell M. C., Oxford.

LaRue, T. A. G., and Kurz, W. G. W. (1973). Estimation of nitrogenase in intact legumes. *Can. J. Microbiol. 19,* 304-305.

Ledoux, L., and Huart, R. (1969). Fate of exogenous bacterial deoxyribonucleic acids in barley seedlings. *J. Mol. Biol. 43,* 243-262.

Ledoux, L., Huart, R., and Jacobs, M. (1974). DNA-mediated genetic correction of thiamineless *Arabidopsis thaliana. Nature 249,* 17-21.

Lie, T. A. (1974). Environmental effects on nodulation and symbiotic nitrogen fixation. In *Frontiers of Biology. Vol. 33. The Biology of Nitrogen Fixation,* (A. Quispel, ed.), pp. 555-582. North-Holland Publishing Co., Amsterdam.

Minchin, F. R. and Pate, J. S. (1973). The carbon balance of a legume and the functional economy of its root nodules. *J. Exp. Bot. 24,* 259-271.

Ohyama, K., Gamborg, O. L. and Miller, R. A. (1972). Uptake of exogenous DNA by plant protoplasts. *Can. J. Bot. 50,* 2077-2080.

Sprent, J. L. (1971). The effect of water stress on nitrogen-fixing root nodules. I. Effects on the physiology of detached soybean nodules. *New Phytol. 70,* 9-17.

Stroun, M., Anker, Ph., and Ledoux, L. (1967). DNA replication in *Solanum lycopersicum* esc. after absorption of bacterial DNA. *Currents in Modern Biology, 1,* 231-234.

Vincent, J. M. (1974). Root-nodule symbioses with *Rhizobium.* In *Frontiers of Biology, Vol. 33, The Biology of Nitrogen Fixation,* (A. Quispel, ed.), pp. 265-341. North-Holland Publishing Co., Amsterdam.

Yao, P. Y., and Vincent, J. M. (1969). Host specificity in the root hair "curling factor" of *Rhizobium spp. Aust. J. Biol. Sci. 22,* 413-423.

New Genetic Approaches to Plant Protection Against Diseases

P. R. Day

During the last 70 years many important crop plant diseases caused by viruses, bacteria, fungi and nematodes have been controlled by using disease resistant cultivars. Today nearly all plant breeding programs are concerned, at least in part, with securing resistance to major diseases. Two forms of resistance are employed: specific resistance (controlled by single genes and highly effective against some pathogen races while not at all effective against others) and general resistance (polygenically controlled and moderately effective against all races of a pathogen) (Day, 1974; Van der Plank, 1975).

Crops protected by a succession of specific resistance genes, replaced in turn to counteract changes in pathogen virulence gene frequencies, depend on a continuing supply of genes. The supply could be conserved by recycling old outmoded resistance genes if the virulence genes that rendered them valueless have disappeared (Person, 1966). In practice, specific resistance genes are often used several at a time in the same cultivar. An alternative is to deploy even more genes in the form of multilines. A multiline variety is made up of 6 to 12 or more components, each with a different resistance gene introduced into a common background cultivar by backcrossing (Browning and Frey, 1969).

Some breeders have abandoned the use of specific resistance in favor of general resistance in spite of the fact that general resistance's polygenic inheritance complicates breeding programs. Clearly, in those crops where general resistance is either inadequate or not yet available, other methods of supplementing specific resistance are necessary.

Another important reason for concern stems from the inherent tendency in modern agriculture towards genetic uniformity (Horsfall 1972). Only the highest yielding, most resistant, most profitable cultivars are grown. This uniformity robs very large land areas of variation to meet unexpected and potentially catastrophic epiphytotics. Variation is a form of insurance that may mean that not every plant or every field will be destroyed. The problem is compounded because no plant breeder can tell which genes will be needed to protect his crop against a new pest or disease that may appear in the years ahead. Protection against two modern epiphytotics, Victoria blight of oats *(Helminthosporium victoriae*) and southern corn leaf blight (*H. maydis* race T), came from our ability to rapidly replace widely grown, uniformly susceptible, cultivars with resistant forms.

Disenchantment with pesticides, the realization that the supply of resistance genes may not be inexhaustible, and the need to rapidly protect otherwise vulnerable crops combine to make the discussion of new genetic approaches to plant protection timely and appropriate. I shall discuss two approaches. The first is through manipulation of pathogens. The second is through work on cultured cells of the host plant.

PATHOGEN MANIPULATION

In the fungi there are several examples of suppressive systems in which cytoplasmic elements invade healthy mycelia through points of hyphal anastomosis and bring about progressive changes or even death. (Fincham and Day, 1973). Some of these systems occur in pathogenic fungi and their relevance to disease control has been explored (Table 3.11 in Day, 1974). Two examples are of interest here.

Chestnut Blight.

Biraghi (1953) recovered in Italy a variant of the pathogen *Endothia parasitica* from stands of European chestnut *(Castanea sativa)* in which blight disease was no longer spreading. The variant was almost avirulent toward inoculated trees. It was subsequently found in France by Grente and Sauret (1969) who called it hypovirulent. They showed that not only does it survive in nature but that it prevents virulent forms from invading host tissue when present in the same canker. When first tested in Connecticut a French hypovirulent strain limited the growth of a virulent French isolate on American chestnut *(C.dentata)* but did not limit canker growth of local indigenous strains of *Endothia* (Anagnostakis and Jaynes, 1973). Isolations made from cankers formed by local strains that were co-inoculated with the hypovirulent strain however, yielded a new form of hypovirulent strain that was effective in restricting invasion of host tissue by local North American strains of *Endothia.* Heterokaryons between the American hypovirulent strain and the American virulent strain (both carrying auxotrophic markers) showed that hypovirulence behaved like a cytoplasmic determininant transmitted by hyphal anastomosis (Van Alfen *et al.*, 1975).

The failure of the French hypovirulent strain to restrict growth of American virulent strains at the first attempt was probably due to heterokaryon incompatibility. Heterokaryon incompatibility restricts exchange of nuclei and cytoplasm to individuals with common alleles at one or several controlling loci and probably functions chiefly to protect healthy mycelia from cytoplasmic infection (Day, 1970; Caten, 1972).

Tests are now in progress to see if the newly derived American hypovirulent strain is persistent in Connecticut and if it will spread and function as a biological control for chestnut blight. The nature of the cytoplasmic determinant of hypovirulence is still unknown.

Corn Smut.

In the course of a survey of North American isolates of *Ustilago maydis,* Puhalla (1968) discovered a cytoplasmically determined trait that he called killer. Haploid sporidial cells of killer strain Pl excreted a protein that inhibited the growth of senstive P2 cells (Hankin and Puhalla, 1971). Crosses between P1 and P2 were performed in corn seedlings and the progeny consisted mainly of P1, a few P2 and a few of another class called P3(1) which was resistant to killer but did not itself produce any killer. In crosses with P2 the resistance of P3(1) also behaved as though cytoplasmically determined. The determinants of P1 and P3(1) were transmitted at heterokaryon formation (Day and Anagnostakis, 1973). Both phenotypes were associated with the presence of virus-like-particles (VLPs) approximately 41 nm in diameter that contained double stranded RNA (dsRNA) which is characteristic of many mycoviruses (Wood and Bozarth, 1973).

In the same survey Puhalla (unpublished) recovered two other killer strains, P4 and P6, with different specificities. P1, P4, and P6 were sensitive to each others' killer proteins. Like P1, P4 and P6 were cytoplasmically inherited, gave rise to resistant forms P3(4) and P3(6) in crosses with P2, and carried VLPs approximately 41 nm in diameter (Koltin and Day, 1976). All three killers, and all three resistant forms contained characteristic dsRNAs that ranged in molecular weight from 0.46 - 2.9 x 10^6 daltons (Koltin and Day, 1976). In addition to cytoplasmically inherited resistance, all three killers are ineffective on strains carrying specific chromosomal resistance factors $p1^r$, $p4^r$, and $p6^r$, respectively, that are unlinked. The sensitive alleles $p1^s$, $p4^s$, and $p6^s$ are all dominant (Puhalla, unpublished; Koltin and Day, 1976). Since all three killer strains when coinoculated with sexually compatible sensitive strains produced infections which gave rise to smut galls, there was little prospect of using the killer phenomenon to control corn smut. In any case, genes for resistance to the killers were not uncommon in wild smut populations.

Koltin and Day (1975) examined the sensitivity of other microorganisms to the killers. None of 51 bacterial samples from 25 species in 10 genera were sensitive, but 55 specimens of 9 different species of graminicolous smuts were sensitive to all three killer proteins. These smuts included the cereal pathogens *Ustilago avenae* (6), *U. hordei* (16), *U. kolleri* (4) and *U. tritici* (12) (number of isolates tested are given in parentheses). A single isolate of a tenth species, *U. sphaerogena,* was slightly inhibited by P1 and P4 and insensitive to P6. In contrast, eight isolates of four other smuts (including *U. violacea* (5)) with dicotyledonous hosts were insensitive except for single isolates of *U. scabiosae* and *U. utriculosa* which were slightly sensitive and sensitive, respectively, to P4. All of the other smuts were tested for ability to inhibit a P2 strain of *U. maydis* sensitive to all three killers. None were inhibitory.

Little is known of either the mode of action or synthesis of the killer proteins. The killer protein may well be coded for by the dsRNA present in the VLPs. If the VLPs can be introduced into cereal plants, either by infecting protoplasts or by insect transmission following insect injection, and if they replicate and produce killer protein in the host, the VLPs could confer resistance to sensitive smuts.

First attempts to infect protoplasts of *U. maydis* strains with dsRNA preparations from P4 have been partially successful (Koltin and Day, unpublished). Normal tobacco tissue is not inhibitory to sensitive cells of *U. maydis* and first attempts to introduce dsRNA or VLPs will be with tobacco protoplasts. Success in this system would encourage us to proceed with cereals.

Mycoviruses appear to be common in the fungi (Bozarth, 1972; Lemke and Nash, 1974) but detailed genetic analysis of their effects on host phenotypes are so far limited to the killer systems of *U. maydis* and a similar system in *Saccharomyces cerevisiae* (Bevan *et al.,* 1973; Vodkin *et al.,* 1974). The three killers of *U. maydis* and the single killer known in yeast may be analagous to the colicins produced by *Escherichia coli* bacteria. Some bacterial DNA plasmids, called colicinogenic factors, infect cells causing them to produce colicins able to inhibit the growth of sensitive bacteria (Reeves, 1972).

CULTURED HOST CELLS

Experimentally induced mutation to resistance is at first sight an attractive alternative to breeding for resistance using naturally occuring genes in crops grown from seed. Starting with a well adapted existing cultivar, only the breeding required to remove unwanted variation and to produce a true breeding homozygote should be necessary. In fact, the results of efforts to induce resistant mutants in crop plants are disappointing. Few, if any, crop cultivars are presently grown that were produced in this way (Day, 1974). One probable reason for this is because disease resistance is often complex and the likelihood of a point change on a chromosome conferring resistance is very small. Some mutants, such as *ml-o* in barley selected for resistance to *Erysiphe graminis hordei,* produced unacceptable necrotic flecks even in the absence of the pathogen. Most studies showed that it was necessary to screen large numbers (10^5 to 10^8) of the treated plants and also to have a reliable and uniform method of distinguishing resistant from susceptible individuals. If the pathogen produces a host specific toxin, the identification problem is relatively simple. Large numbers of plants can be handled by using young seedlings and dosage can be controlled by adjusting toxin concentrations to kill susceptible plants, allowing only resistant plants to survive (Day, 1974).

Cultured host cells enable very large numbers to be screened, the only restriction being that for practical use one must be able to regenerate whole plants from survivors. Carlson (1973) was the first to use this technique on mutagenized haploid cells of *Nicotiana tabacum. Pseudomonas tabaci* is a bacterial pathogen that causes wildfire disease of tobacco. Cells resistant to wildfire toxin and to methionine sulfoximine, an analog of the toxin, were selected. Three diploid plants regenerated from methionine sulfoximine resistant calluses were significantly less susceptible to the pathogenic effects of bacterial infection than the parent plants. Their leaves were resistant to the toxin produced by the pathogen but did not restrict pathogen development. In two of the plants, resistance was transmitted in crosses as a single semidominant character while in the third it was more complex.

More recently, Gengenbach and Green (1975) have exposed diploid cells of a corn line that carried the *cms-T* factor for cytoplasmic male sterility to toxin. Plants and plant parts (including cultured cells) that carried this cytoplasmic factor were susceptible to the toxin(s) that accumulated in culture filtrates of race T of *Helminthosporium maydis.* Normal male-fertile plants and tissues that lacked *cms-T* were resistant to the toxin. Race T was responsible for the epiphytotic of southern corn leaf blight that swept the corn growing regions of North America in 1970 (Ullstrup, 1972).

After four subculture cycles of selecting the fastest growing clones on media containing sublethal toxin concentrations, Gengenbach and Green recovered some 14 callus clones that grew on toxin concentrations that were lethal to unselected callus. The selected clones were stable and still resistant when retested after more than four months growth in the absence of toxin. The toxin induced changes in membranes of mitochondria isolated from *cms-T* cells (Gengenbach *et al.,* 1973) and also uncoupled and inhibited mitochondrial electron transport (Peterson *et al*., 1975). Mitochondria isolated from resistant callus, like those from normal tissue, were not sensitive to toxin.

At the present time it is not known whether plants can be regenerated from the resistant callus and, if so, whether they would be male fertile or male sterile. The resistant clones may indeed have arisen by mutation of either nuclear or cytoplasmic DNA. It is also possible, however, that *cms-T* plants contained small numbers of normal mitochondria and that the few cells which retained them were selected during exposure to toxin. This theory supposes that the *cms-T* determinant is part of the mitochondrial DNA. Another explanation may be that a cytoplasmic determinant of *cms-T,* independent of the mitochondria, may sometimes fail to be included in all cells and that cells without it can form normal mitochondria, are toxin resistant, and thus selected. Both selection hypotheses suggest that normal cells may arise in whole *cms-T* plants. Whether they would give rise to detectable chimeras would depend on whether reinfection occurs via plasmodesmata.

Several other examples of the use of pathogen toxins to select resistant mutants as seedlings include Victoria oats resistant to *Helminthosporium victoriae* (Luke and Wallace, 1969), and sorghum reisitant to *Periconia circinata* (Schertz and Tai, 1969). Sugarcane seedlings susceptible to eyespot *(H. sacchari)* are now eliminated by treatment with the toxin helminthosporoside early in breeding (Steiner *et al.*, 1972). Strobel (1973) reported that susceptible sugarcane clones possess a plasma membrane component which is a toxin binding protein that is not present in resistant clones. Recently he was able to show that resistant sugarcane leaf cell protoplasts and tobacco protoplasts, both insensitive to the toxin, took up this isolated binding protein and as a result became susceptible to helminthosporoside (Strobel, 1975). This demonstration supports his hypothesis on the nature of eyespot resistance and susceptibility; namely that binding of helminthosporoside alters the membrane to cause electrolyte imbalance and cell death. Presumably, the induced susceptibility caused by taking up the binding protein is temporary and would be diluted out either by growth or membrane turnover if protoplast regeneration were attempted.

In a recent review, Scheffer and Yoder (1972) listed some eight examples of fungal pathogens that produce specific toxins. Durbin (1972) and Patil (1974) have also discussed the toxins of phytopathogenic bacteria which, although not host specific to the extent of some fungal toxins, can, as Carlson (1973) showed, be used to select resistant mutants.

What are the prospects for using these methods to select for resistance to other pathogens that do not produce toxins? Although several authors have found that cultured tissues of resistant and susceptible cultivars can be distinguished by their responses to pathogen inoculum (Ingram, 1973), the results are not promising. For example, with *Phytophthora parasitica* var. *nicotianae*, Helgeson and co-workers (1972) found that relatively small changes in temperature, inoculum density, callus morphology, and hormone content of the medium all had profound effects on pathogen colonization of cultured tobacco tissues. Resistant and susceptible interactions were distinguished by their different rates of colonization. While such studies may be useful for examining the nature of resistance, they do not give the sharp discrimination required for positive selection of very infrequent resistant mutants.

Facultative parasites frequently grow so well on media supporting callus growth that they overwhelm the plant tissue. Where pathogen growth can be controlled, the method offers the opportunity to look for inhibition zones surrounding callus pieces that may produce toxic materials.

Our knowledge of resistance mechanisms in crop plants and the use we make of them is still largely empirical. When we better understand how resistance works and especially why it often fails, we should be able to devise better methods for selecting new forms of resistance.

There is much interest in novel methods for transcending the bounds imposed by sexual compatibility of exchanging genetic information between eucaryotes. Interspecific and intergeneric sexual crosses have been a valuable means of introducing alien chromosomes and the genes for resistance they carry. If protoplast fusion and methods using potential DNA vectors, such as cauliflower mosaic virus or *Agrobacterium tumefaciens,* can be developed to the point of success there could arise the prospect of employing 'non-host resistance' in breeding programs.

REFERENCES

Anagnostakis, S. L., and Jaynes, R. A. (1973). Chestnut blight control: use of hypovirulent cultures. *Plant Dis. Reptr. 57,* 225.

Bevan, F. A., Herring, A. J., and Mitchell, D. J. (1973). Preliminary characterization of two species of dsRNA in yeast and their relationship to the killer character. *Nature 245,* 81-86.

Biraghi, A. (1953). Possible active resistance to *Endothia parasitica in Castanea sativa. Report 11th Congress Int. Union for Res. Org.,* 643-645.

Bozarth, R. F. (1972). Mycoviruses: a new dimension in microbiology. *Environmental Health Perspectives 2,* 23-29.

Browning, J. A., and Frey, K. J. (1969). Multiline cultivars as a means of disease control. *Ann. Rev. Phytopath 7,* 355-382.

Carlson, P. S. (1973) Methionine-sulfoximine-resistant mutants of tobacco. *Science 180,* 1366-1368.

Caten, C. E. (1972). Vegetative incompatibility and cytoplasmic infection in fungi. *J. Gen. Microbiol. 72,* 221-229.

Day, P. R. (1970). The significance of genetic mechanisms in soil fungi. In *Root Diseases and Soil-borne Pathogens* (T. A. Tousson, R. V. Bega and P. E. Nelson, eds), pp. 69-74. Univ. Calif. Press, Berkeley.

Day, P. R. (1974). *Genetics of Host-parasite Interaction.* Freeman, San Francisco.

Day, P. R., and Anagnostakis, S. L. (1973). The killer system in *Ustilago maydis:* heterokaryon transfer and loss of determinants. *Phytopath. 63,* 1017-1018.

Durbin, R. D. (1972). Bacterial phytotoxins: a survey of occurrence, mode of action and composition. In *Phytoxins in Plant Disease* (R. K. S. Wood, A. Ballio and A. Graniti, eds.), pp. 19-33. Academic Press, London.

Fincham, J. R. S. and Day, P. R. (1973). *Fungal Genetics. 3rd Edit.* Blackwell, Oxford.

Gengenbach, B. G. and Green, C. E. (1975). Selection of T-cytoplasm maize callus cultures resistant to *Helminthosporium maydis* race T pathotoxin. *Crop Sci.* 15, 645-649.

Gengenbach, B. G. Miller, R. S., Koeppe, D. E., and Arutzen, C. J. (1973). The effect of toxin from *Helminthosporium maydis* (race T) on isolated corn mitochondria swelling. *Can. J. Bot. 51,* 2119-2121.

Grente, J., and Sauret, S. (1969). L. hypovirulence exclusive, phenomene original en pathologie vegetale,*C. R. Acad. Sc. Serie D, 268, 2347-2350.*

Hankin, L., and Puhalla, J. E. (1971). Nature of a factor causing interstrain lethality in *Ustilago maydis*. *Phytopath. 61,* 50-53.

Helgeson, J. P., Kemp, J. D., Haberlach, G. T., and Maxwell, D. P. (1972). A tissue culture system for studying disease resistance: the black shank disease in tobacco callus cultures. *Phytopath 62,* 1439-1443.

Ingram, D. S. (1973). Growth of plant parasites in tissue culture. In *Plant Tissue and Cell Culture* (H. E. Street, ed.) pp. 392-421. Univ. Calif. Press, Berkeley.

Horsfall, J. G. (1972). *Genetic Vulnerability of Major Crops.* Nat. Acad. Sci., Washington.

Koltin, Y., and Day, P. R. (1975). Specificity of *Ustilago maydis* killer proteins. *Applied Microbiol. 30,* 694-696.

Koltin, Y., and Day, P. R. (1976). Inheritance of killer phenotypes and double-stranded RNA in *Ustilago maydis. Proc. Nat. Acad. Sci. (U.S.A.) 73,* 594-598.

Lemke, P. A., and Nash, C. H. (1974). Fungal viruses. *Bacteriol. Rev. 38*, 29-56.

Luke, H. H., and Wallace, A. T. (1969). Sensitivity of induced mutants of an *Avena* cultivar to victorin at different temperatures. *Phytopath. 59,* 1769-1770.

Patil, S. S. (1974). Toxins produced by phytopathogenic bacteria. *Ann. Rev. Phytopath. 12,* 259-279.

Person, C. (1966). Genetic polymorphism in parasitic systems. *Nature 212,* 266-267.

Peterson, P. A., Flavell, R. B., and Barratt, D.H.P. (1975). Altered mitochondrial membrane activities associated with cytoplasmically-inherited disease sensitivity in maize. *Theoret, Appl. Genet. 45,* 309-314.

Puhalla, J. E. (1968). Compatibility reactions on solid medium and interstrain inhibition in *Ustilago maydis. Genetics 60,* 461-474.

Reeves. P. (1972). *The Bacteriocins.* Springer-Verlag, New York.

Scheffer. R. P., and Yoder, O. C. (1972). Host-specific toxins and selective toxicity. In *Phytotoxins in Plant Diseases* (R. K. S. Wood, A. Ballio and A. Graniti, eds.), pp. 251-272. Academic Press, London,

Schertz, K. F., and Tai, Y. P. (1969). Inheritance of reaction of *Sorghum bicolor* (L.) Moench to toxin produced by *Periconia circinata* (Mang) Sacc. *Crop. Sci. 9,* 621-624.

Steiner, G. W., Byther, R. S., Strobel, G. A., and Hess, W. M. (1972). *Hawaiian Sugar Planters Association 1971 Annual Report* , p. 41.

Strobel, G. A. (1973). Biochemical basis of the resistance of sugarcane to eyespot disease. *Proc. Nat. Acad. Sci. (U.S.A.) 70,* 1693-1696.

Strobel, G. A. (1975). Transfer of toxin susceptibility to plant protoplasts via the helminthosporoside binding protein of sugarcane. *Biochem. Biophys. Res. Comm. 63,* 1151-1156.

Ullstrup, A. J. (1972). The impacts of the southern corn leaf blight epidemics of 1970-1971. *Ann. Rev. Phytopath. 10*, 37-50.

Van Alfen, N. K., Jaynes, R. A., Anagnostakis, S. L., and Day, P. R. (1975). Chestnut blight: biological control by transmissible hypovirulence in *Endothia parasitica. Science 189,* 890-891.

Van der Plank, J. E. (1975). *Principles of Plant Infection.* Academic Press, New York.

Vodkin, M., Katterman, F., and Fink, G. R. (1974). Yeast killer mutants with altered double-stranded ribonucleic acid. *J. Bacteriol. 117,* 681-686.

Wood, H. A., and Bozarth, R. F. (1973). Heterokaryon transfer of viruslike particles associated with a cytoplasmically inherited determinant in *Ustilago maydis. Phytopath. 63,* 1019-1021.

Index

D

E

F

G

H

I, K

L

M

N

O

P

R

S

T

U

V

W, Z

A
B 7
C 8
D 9
E 0
F 1
G 2
H 3
I 4
J 5